Engraved by A.H. Ritchie.

WILLIAM TAYLOR.

SEVEN YEARS'

STREET PREACHING

IN

SAN FRANCISCO, CALIFORNIA;

EMBRACING

INCIDENTS, TRIUMPHANT DEATH SCENES, ETC.

BY REV. WILLIAM TAYLOR,

OF THE CALIFORNIA CONFERENCE.

"Go out quickly into the streets and lanes of the city."
"The common people heard him gladly."

EDITED BY W. P. STRICKLAND.

THIRTY-SECOND THOUSAND.

New York:

PUBLISHED FOR THE AUTHOR

BY CARLTON & PORTER, 200 MULBERRY-STREET.

INTRODUCTION.

There is a certain class of books which need no introduction, a single glance at their title and contents being sufficient to secure them a reading. It is unnecessary to say that the following work is of that character. The author, as various allusions in the body of the work will show, was a street preacher before he went to California, and in the streets of the Monumental City and the Capital he received that initiatory training and qualification for addressing crowds in the open air, which fitted him so successfully for preaching the Gospel among the crowding thousands on the shores of the Pacific, before the erection of churches. His labors at home in this particular department of Christian effort doubtless pointed him out to the authorities of the Church as the very man for that new and interesting field, so strangely opened by Providence to the citizens of the United States.

In the fall of 1848, while passing up Baltimore-street, he heard some one call his name. On turning in the direction of the sound, he saw Christian Keener running toward him, exclaiming, "Bishop Waugh wants to see you, at Armstrong & Berry's book-store." He accordingly returned, and the bishop informed him, that if he had no objections, he would send him as a missionary to California. Always ready to obey the voice of the Church, he responded to the call, and, with his family, was in due time on his way to that distant field.

The reader will find, in perusing these pages, the results of his seven years' labors in the land of gold and crime; and as he follows him from street to street, from hospital to Bethel, and from plaza to quay, and hears his songs and sermons, and reads his sketches of men and manners, and the various scenes through which he passed, will learn more of the real social and political condition of the country than has ever yet been furnished by books or journals.

The following letter, from the Hon. Wilson Flint, Senator in the California State Legislature, which was sent unsolicited, will show to what extent Mr. Taylor's labors were successful, as well as the manner in which they were appreciated:

San Francisco, Sept. 10, 1856.

Rev. Wm. Taylor:

My esteemed Friend—Learning that you were intending to publish a work, containing a history of your labors as a preacher of the Christian religion at San Francisco, from your arrival here in 1849 to the present time, I thought it would be in place for me to remind you of an incident which occurred in the Plaza, the first time I had the pleasure of listening to you.

It was on a Sunday morning, in December, 1849, when landing from the Panama steamer, I wended my way with the throng to Portsmouth Square, this being at that time the great resort of the denizens of the rising metropolis. Three sides of the square were mostly occupied by buildings, which served the double purpose of hotels and gambling-houses, the latter calling being regarded at that time as a very reputable profession. On the fourth and upper side of the square was an adobe building, from the steps of which you were discoursing from the text, "The way of the transgressor is hard."

It was a scene I shall never forget. On all sides of you were gambling-houses, each with its band of music in full blast. Crowds were going in and out; fortunes were being lost and won, terrible imprecations and blasphemies rose amid the horrid wail, and it seemed to me that Pandemonium was let loose. Above all this, I heard you utter the following prophetic sentence, which has since been fully realized: "The power of Satan seems at this time in the ascendency, wherever I cast my eye; but, sure as there is a God in heaven, we will turn the tables upon the Evil One, and where now my voice meets naught but scoffs and jeers, with unwavering faith in my Divine Master, I hope to labor on to the time when these dens of iniquity around me shall all be swept away."

Six years of time have sped on, and what a wondrous change! Portsmouth Square now, of a Sabbath morn, is thronged with women and children wending their way to the numerous churches in the surrounding localities. A great metropolis spreads out on every side, and civilization and Christianity go hand in hand to humanize the race of man.

Yours obediently,

WILSON FLINT.

Numerous as have been the works published in relation to California, its religious history has not been written; and hence the present work occupies a department altogether new in California literature. As such, we have no doubt it will be sought after with avidity, and read by thousands. The work also supplies a desideratum in the literature of the Church, being a most admirable treatise on open-air preaching, illustrated by the labors of ten years. Its incidents and triumphant death scenes are of a most thrillingly interesting character, and the latter exemplify, in an eminent degree, the excellence and value of religion.

EDITOR.

NEW-YORK December 19, 1856.

Illustrations.

CONTENTS.

SEVEN YEARS' STREET PREACHING.

CHAPTER I.

FIRST PREACHING ON THE PLAZA.

On the third of December, 1849, I announced to the congregation in "Our little church on the hill," that at three P. M. of that day I would preach in the open air, on Portsmouth Square, known more familiarly as the Plaza. It was regarded by most persons present, if not all, as a very dangerous experiment; for the gamblers were a powerful and influential party in the city, and the Plaza was their principal rendezvous, and Sunday the best day of the seven for their business. The Plaza was nearly surrounded by gambling and drinking houses. The gamblers occupied the best houses in the city, and had them furnished in the most magnificent style. Each house employed a band of the best music the country could afford. R. Beeching, a member of our Church, being a good musician, was offered thirty dollars per night to play in one of them, which, as a true man, though poor and out of employment at the time, he declined.

The walls of these houses were hung with splendid

paintings; "the tables" contained "piles" of gold and silver; the musicians occupied a high platform in the rear end of the saloon; the "needful" was served out by "a gentleman of the bar," in one corner, near the entrance, where many a jolly circle drank to each other's health the deadly draught. These places, especially at night, all night, and on Sunday, were crowded with moving masses of humanity, of every age and complexion. So powerful was this class of men in the city, that I do not remember of ever hearing of one of them, in those days, being arrested, even for murder. Now, should a poor preacher presume to go into their midst, and interfere with their business, by thrilling every house with the songs of Zion and the peals of Gospel truth, he would be likely to wake up the lion in his lair! When the appointed hour arrived I took with me my "sweet singer in Israel," the partner of my youth, who has stood by me in every battle; and down I went to the field of action. I selected for my pulpit a carpenter's work-bench, which stood in front of one of the largest gambling-houses in the city. I got Mrs. T. and another lady or two comfortably seated, in care of a good brother, and taking the stand, I sung on a high key,

"Hear the royal proclamation,
The glad tidings of salvation,

Publishing to every creature,
To the ruin'd sons of nature,
Jesus reigns, he reigns victorious
Over heaven and earth most glorious.
"Jesus reigns," etc.

The novelty of the thing had a moving effect. The people crowded out of the gambling-houses, and gathered together from every direction, as though they had heard the cry, "Fire! fire! fire!" By the time the echoes of the song had died on the breeze, I was surrounded by a dense crowd, to whom I introduced the object of my mission, as follows: "Gentlemen, if our friends in the Atlantic states, with the views and feelings they entertained of California society when I left there, had heard that there was to be preaching this afternoon on Portsmouth Square, in San Francisco, they would have predicted disorder, confusion, and riot; but we, who are here, believe very differently. One thing is certain, there is no man who loves to see those stars and stripes floating on the breeze, (pointing to the flag of our Union,) and who loves the institutions fostered under them; in a word, there's no true American but will observe order under the preaching of God's word anywhere, and maintain it, if need be. We shall have order, gentlemen. I apprehend that for the last twelve months at least, you have all been figuring under the rule of 'loss and gain.' In your tedious voyage 'round the Horn,' or your wearisome journey over the Plains, or your hurried passage 'across the Isthmus,' and during the few months of your sojourn in California, losses and gains have constituted the theme of your thoughts and calculations. Now, I wish most respectfully to submit to you a question under your favorite rule. I want you to employ all the mathematical power and skill you can command, and patiently work out the mighty problem. The question may be found in the twenty-sixth verse of the sixteenth chapter of our Lord's Gospel by St. Matthew. Shall I announce it? 'What is a man profited, if he shall gain the whole world, and lose his own soul?'"

Every man present was for that hour "a true American." Perfect order was observed, and profound attention given to every sentence of the sermon that followed. The warrant for street preaching in San Francisco was thus acknowledged, and the precedent of good order, under the preaching of the word in these "highways," was thus established. That sermon proved to be the first of a series of nearly six hundred sermons preached in these streets, the confluence of all the various creeds, and isms, and notions, and feelings, and prejudices of the representatives of all the nations, Christian and heathen. And yet, through the restraining providence of Him who sent me, and the good common sense of the people of

California, I have never lost a congregation, nor suffered any serious disturbance. The little interruptions I have had, together with specimen extracts from sermons preached, and incidents illustrating some of the modes by which truth has been presented, will be duly noted in the subsequent pages of this book.

CHAPTER II.

STREET PREACHING.

"Why do you preach in the streets and highways?" I answer,

I. Because it is a duty enjoined by the Lord Jesus Christ.

The "great commission," under which every true embassador goes forth in the "ministry of reconciliation," by direct *implication*, enjoins the duty of out-door preaching: "*Go ye into all the world, and preach the Gospel to every creature.*" Did the apostles understand the Great Teacher to mean that they were to preach in the temple, in the synagogues, in "hired houses," and "upper rooms?" Certainly. Did they understand him to mean nothing more than that? Certainly not. They well knew that the temple, and the synagogues, and all the house room they could by possibility command, were they all open for their use, would contain but a very small proportion of the creatures embraced in their commission. Every word of this great command, framed by infinite wisdom, is simple and unequivocal. It evidently contemplates a proclamation of the Gospel as wide as "all out of doors," and so specific and personal as to embrace every single rebel of the fallen race.

Again. The Saviour, illustrating, by the parable of the "Great Supper," the bounteous provision of mercy in the Gospel, enjoins, by *direct command*, the duty of out-door preaching: "Go out quickly into the streets and lanes of the city, and bring in hither the poor, and the maimed, and the halt, and the blind." "Go out into the highways and hedges, and compel them to come in, that my house may be filled."

II. It is supported by Divine and apostolic precedent and example.

The only sermon of our Divine teacher on record, was preached on a mountain. Many others of which we have full "reports," were preached by the sea-shore, on the decks of ships, and in the streets of Capernaum. He preached, to be sure, in the temple and in the synagogues, but of his sermons on those occasions, there is less recorded than of his "out-door sermons." We believe that he established, by his own example, the *precedents* he designed to be practically operative through all time, namely: To get all we can into the synagogues and churches, and there preach to them, and then to "go out into the streets and lanes of the cities, and into the highways and hedges," and hunt up *all the rest*, and preach to them also. The apostles acted accordingly. The great apostle to the Gentiles was celebrated as an out-door preacher.

III. It has been confirmed by a Divine attestation, in that God has always *signally* owned and blessed the faithful preaching of the word in the streets, lanes, highways, and hedges, through all the history of the Church to the present time. Without going back to instance the singular courage displayed and success attained by the out-door preaching of some of the Vaudois missionaries, in the Dark Ages, we would invite attention to the "field and street preaching" of more modern days. Witness the labors and successes of Whitefield, and Wesley, and Fletcher, and their coadjutors in the same work. God made the out-door preaching of those men a leading instrumentality in awaking the masses of the "United Kingdom of Great Britain," and in bringing about the great reformation of the eighteenth century. See the labors and good fruits of the street preaching of "the apostle to the Irish," Gideon Ouseley. Should there be a resurrection of the brute creation, the experience of "Ouseley's white horse," from whose back he so often preached, will furnish an interesting chapter in the annals of "the new heavens and new earth." Again, witness, in quite modern times, the street preaching and Gospel triumphs of the champions of the "Free Church of Scotland."

Thus Jesse Lee drove the entering wedge of Methodistic Christianity into New-England. American camp-meetings of the different Churches come under the apostolic precedent of out-door preaching. See how they have been honored of God. Recount, if you can, the multiplied thousands of souls who have been converted at camp-meetings, multitudes of whom have washed their robes in the blood of Jesus, and are to-day seated above the circle of the heavens, praising God for camp-meetings. Strike out of the Church in America all her ministers and members who have been brought to God through the instrumentality of camp-meetings, and you will have a practical demonstration of the truth of our position, which will astonish you. I instance camp-meetings not as a proof that the Gospel ministers of America have fulfilled all their duty in regard to out-door preaching, but as evidence that they have gifts eminently qualifying them for that work, and especially to demonstrate the truth of the position with which I set out, namely, that God has always signally owned and blessed the out-door preaching of his embassadors. But says one, " The most of your instances relate to periods when houses of worship were not available, or were entirely inadequate to meet the demand. We have plenty of good churches now, and if the people want to hear the Gospel, let them come to church." Thank you, sir; that will help me to another argument, which I will, according to "Thomsonian practice," call Number

IV. Demonstrating the moral necessity of street preaching, on the principle suggested, namely: *The facts as they are exhibited in the present history of the world.*

Passing by heathendom and foreign Christian countries, I will confine my investigations to our own country. And now allow me to inquire of the objector, What proportion of the population of your town or city will your churches accommodate? And what proportion of the people attend church? Now, what can you do for those "creatures" embraced in the provisions of your Divine commission, but not embraced in your church accommodations? To say that the vast proportion of your non-church-going population must either come to Church or go to hell unwarned, is to institute a new condition of salvation unknown to the Gospel. Now, in California, all the churches, Catholic and Protestant, will accommodate, say sixty thousand persons, which will leave two hundred and forty thousand "outsiders," and no church room to receive them; "no, not so much as about the door." What would Jesus do in such a case? There is room enough for every one of them in the compassion of God, and in the kingdom of grace; for they "are not common or unclean," not excluded from the covenant of promise. Very many of them are the sons of our fathers and mothers in Israel, who have died in the faith. They went down to their graves praying for their children. The last words they uttered, as one by one they left the shore, were, "Tell my dear children to meet me in heaven?" Their sons have become prodigal and reckless in California, and yet the mention of their sainted mother checks the giddy laugh, and brings tears to their eyes. What would those fathers and mothers have us do for their wandering, lost children in the wilderness? Would they not say, would not the angels say, as Jesus hath said, once for all, "Go out quickly, and compel them to come in?" And do you not respond, "Amen! Go, my brother; go out quickly?"

Though I have been singing and preaching the "royal proclamation," in the "highways," to these wanderers, for seven years, my tears would now, as I write, saturate the manuscript, at the remembrance that I have done so little to save them, and that I have seen so many hundreds of them dying in this land, without any hope of heaven.

But let me ascend to a higher stand-point, and take a wider view of the subject. According to statistics furnished by the United States Census of 1850, the Methodists of this great republic have 12,467 churches, which would accommodate 4,209,333 persons. Now, we profess to "believe that God's design, in raising up the people called Methodists, was to reform the continent, and spread Scriptural holiness over these lands." I am no croaker. I think I have a just appreciation of the great work God hath wrought through the Methodists, and other denominations of Christians as well, and I think I am unfeignedly thankful. But let us look again at the facts before mentioned. We have been engaged in this work of reforming the continent for upward of eighty years. We have in our favor the constitution and laws of the land, one common language, ready access to all classes of society, and every desirable facility for communicating truth. We also have at command the mighty appliances of our mighty Gospel, and the spiritual resources of Omnipotent grace. And yet, in all this lapse of four-score years, we have only reached with the sound of the Gospel jubilee about one fifth of the population of the United States. As embassadors of Christ, we have, at last, "challenged for a hearing" say 5,000,000 at one time, a little over 1,500,000 of whom, now living, have closed with the terms of the Gospel, and are now reconciled to God. The remaining 3,500,000 have possibly taken the matter under advisement, leaving say 20,000,000 of precious souls for whom we have no room in our churches. What shall we do to reach them, especially the masses not embraced under the ministry of any other Church?

But I would go up still higher, to a point whence I can have one grand view of the whole "field." According to the census returns of 1850, all the churches of the United States, Catholic, Protestant, and all together, will accommodate 14,000,000, leaving about 10,000,000 of souls in this Christian republic, for whom there is no room in any of the churches. Four tenths of the population of these United States never go to church!

"But," it is asked, "does not a much larger number than that indicated by the aggregate capacity of the churches occasionally attend church, alternately with other occasional church-goers?" If you will make a calculation of the actual aggregate attendance in the churches throughout the land, you will find the number resulting from such a calculation so far below the number indicated by the aggregate capacity of the churches, that you will have plenty of room left to accommodate all your occasional church-goers, without calling on one of our outside ten millions.

Now, it is a question, which rises infinitely above any mere sectarian view of the subject, How shall the enlightening, purifying, elevating influences of the Gospel be brought to bear upon this mighty mass of neglected humanity? "Righteousness exalteth a nation, but sin is a reproach to any people." Our nation is reproached and enfeebled far beyond safe precedent by the church-going sinners, but look down to the lower strata, and see the ten millions who have no fear of God before their eyes, no care for the honor of the nation, no sympathy with the grand institutions which every American citizen should cherish as he does his life. They are, by their accumulating vices, locking the very wheels of government. They are corrupting the life-blood of the body politic. And they are deteriorating rapidly, and multiplying continually; first, by foreign immigration, and, second, by their own children, brought up under the special tutorship of Satan. Among the foreign immigration to our shores, are very many whose citizenship would be an honor to any nation; but a large proportion may be set down, at best, as fifth-rate humanity, morally considered.

These millions of neglectful and neglected souls are all subjects of redeeming mercy in Christ, and the infinite heart of Jesus, with every pulsation, throbs in sympathy with their woes. They can be redeemed and elevated to good citizenship in a

Christian republic, and to heirship in the kingdom of glory. But the question remains, How shall they be reached and saved? The statesman replies, "Educate the masses, multiply public schools, academies, and colleges; teach every prattler in the nation how to read." That is a suggestion worthy of a statesman, a most desirable end to be attained. But will mere intellectual training, however important, secure the end proposed? Hear what General Washington says on this subject: "Of all the dispositions and habits which lead to political prosperity, religion and morality are indispensable supports. In vain would that man claim the tribute of patriotism who should labor to subvert the great pillars of human happiness, those firmest props of men and citizens! The mere politician, equally with the pious man, ought to respect and cherish them. A volume could not trace all their connections with private and public felicity. Let it be simply asked, Where is the security for property, for reputation, for life, if the sense of religious obligation desert the oaths which are instruments of investigation in the courts of justice? And let us indulge with caution the supposition that morality can be maintained without religion. Whatever may be conceded to the influence of refined education on minds of a peculiar structure, reason and experience both forbid us to expect that national morality can prevail in exclusion of religious principle." Educate a rogue, (I use the term "educate" in the popular sense of intellectual training,) and the increase of his intellectual power will but make him the more accomplished as a rogue, and proportionably more dangerous to society. The American Bible Society, the grand Christian institution on whose catholic platform all denominations of Christians meet, and pray, and labor together, and realize and exhibit "how good and how pleasant it is for brethren to dwell together in unity," has pledged itself to place a Bible in the hands of every American family, a conception and purpose worthy so noble an institution. The Tract Societies of different grades are doing a great work for the moral improvement of the masses. An increasing diffusion of religious literature in general is looked upon, and justly too, as a great means of good to society. But, after all, the question arises, Will a man, who never goes to church, nor desires to go, read the Bibles, and tracts, and religious books you put into his hands? A few may, but the mass of such people certainly will not. They have no desire for religion, and no taste for religious literature, and they are not so self-denying as to spend time in reading what is not agreeable to their views and feelings. These means of diffusing religious knowledge, however important as auxiliaries, do not constitute the peculiar instrumentality ordained of God for the enlightenment and salvation of the world. If this were so, then the great commission would have been framed accordingly, and would read: "Go, publish Bibles, tracts, papers, and religious books, and scatter them abroad as 'leaves from the tree of life for the healing of the nations.'" Jesus says no such thing; but "*Go* YE *into all the world, and* PREACH the Gospel TO EVERY CREATURE." Let the embassadors of Jesus use all the collateral appliances at their command, as valuable aids, but not to be substituted for God's appointment of PREACHING *the Gospel.*

The whole matter resolves itself into this, that these ten millions of our neighbors, whom we are commanded to love as we love ourselves, must have "the Gospel preached unto them," or the mass of them will go to perdition. They are blinded by the god of this world, and will not come to us. Should we not, in the name and spirit of Him who came to seek and to save the lost, "go" to them?

In concluding this argument, I would most respectfully submit a suggestion for the consideration of wise and good men. Let a good representation of the American pulpit, for the love of souls, as the visible representatives of Jesus, "go out into the highways, and preach the Gospel." Let each act upon his own responsibility, as he that must give an account; but, as far as practicable, let the ministers of all denominations act in concert. Let them, like the ancient heralds of the great jubilee of the Jews, simultaneously "proclaim liberty throughout all the land, unto all the inhabitants thereof." Let the rising, swelling blasts of ten thousand trumpets, unite their echoes from Dan to Beersheba, from Maine to New Mexico, and from South Carolina to California; and let all the laity, who "know the joyful sound," "run to and fro," bearing the "tidings," personally, to their neighbors, and knowledge shall be increased.

CHAPTER III.

OBJECTIONS TO STREET PREACHING CONSIDERED.

I. Do I HEAR YOU SAY IT IS A DEGRADATION OF MINISTERIAL DIGNITY?

I reply: Any minister of the Gospel, whose "ministerial dignity" depends, for its elevation and support, upon the sacredness of a consecrated pulpit, is not, I confess, a suitable person for a street preacher. A preacher, to succeed in the streets, must be dignified by a special unction of the Holy Spirit. He must feel such an undying thirst for the salvation of sinners as will prompt him, like Aaron, to run out into the camp, and "stand between the living and the dead;" not only to offer the incense of earnest prayer to God on their behalf, but also to warn them from the example of their neighbors, who have perished in their sins. Then the accompanying presence of Him who hath said, "Lo, I am with you always, even unto the end of the world," will consecrate any place in which he may open his commission, as much as the spot where "Jacob slept, and dreamed, and saw the ladder that reached from earth up into heaven;" and cause every one to feel, "Surely the Lord is in this place. How dreadful is this place! This is none other but the house of God; and this is the gate of heaven."

II. It makes the preaching of the Gospel too common.

I think there is much more danger of making the preaching of the Gospel too uncommon than "too common;" too abstract in the matter of it, and too high in its *mode* of delivery. Not common enough to flow readily into the common channels of human thought and sympathy, nor materially to affect the common relations and conduct of men. A man, to succeed as a street preacher, must be eminently practical in his preaching. Nothing but the simple preaching of our common Gospel, in a manner to arrest the attention and engage the feelings of the common people, will enable him to get, or to hold, an audience in the "highways." It was this that made the "common people" hear Jesus gladly. I will here add, that the street is not the place for sectarian discussions; but the Gospel, in all its essential characteristics, should be clearly announced.

III. It will detract from the interest of the people in the regular ministrations of the pulpit.

Such a proposition is philosophically unsound, and is contradicted by the facts in all history relating to this subject. Street preaching, where churches were not, has always led to their erection, and when efficiently administered, even in old cities, has always contributed to increase the congregations in church. Such is, so palpably, the testimony of history, that I need not instance the proofs; and such is the result of my own observations. I had the honor of preaching two years (excepting the cold weather) in the streets of Georgetown, D. C. The effect was manifestly good in the increase of the regular church audiences. And in a revival there, under the superintendence of Rev. Henry Tarring, of precious memory, now in glory, quite a number of the converts testified that they received their religious awakening under the "market-house preaching." Among those converts were several Roman Catholics, who had never heard Protestant preaching until attracted by the street exercises.

I also had the pleasure of preaching a year in Belair Market, Baltimore City. Two persons, I remember, kneeled on the pavement, and cried for mercy, and were there happily converted. One of them, by the name of "Shilling," I learn is still a useful member of the Church in North Baltimore Station. During the summer of my preaching in that market, "Father Darling," the sexton of the Monument-street

Church, who knew the faces of all the regular auditors, said, "I cannot tell where so many strangers come from. They keep coming in every Sunday night, more and more." During the revival in the fall, under the direction of my much-esteemed colleagues, Revs. C. B. Tippett and J. S. Martin, a number of those strangers made a profession of religion, and testified that, though they had lived for years in the city, they had not been to church, till attracted by the market-house preaching. My worthy colleagues there, took a part in the street preaching.

In the city of San Francisco my street preaching has been a regular advertisement for the churches in general, for I always take occasion to announce the church appointments. It has always contributed to our church congregations; and a majority of those whom I had the happiness of seeing converted in this wicked city, say two hundred, testified to the fact that they were awakened under the street preaching. This city, however, does not furnish a fair test of the legitimate effect of the preached word, in doors or out, as I will take occasion to show in the progress of this work.

IV. It creates riots and confusion in the streets.

I apprehend that much of the trouble which has been heard of in connection with street preaching, resulted from injudicious attacks upon Romanism, or upon personal character, or for want of tact in controlling large audiences. I do not know, definitely, the merits of any given case, but can readily see how, in various ways, a man could bring upon himself a great deal of trouble, and defeat the object of his mission.

Still, "men love darkness rather than light," and it would not be surprising if an earnest, faithful, modern street preacher should share the same lot that St. Paul did at Athens, Philippi, and other places, but we never learned that the apostle considered that a sufficient objection to lead him to desist from preaching in the streets. I have been preaching regularly in the streets for more than ten years, and seven of them among California gamblers and rum-sellers, and through the good providence of the Lord, I have never had a serious disturbance, nor lost a congregation in the streets.

V. It will bring the preacher into collision with the civil authorities.

We should be careful, while we do our duty fearlessly, not to provoke a collision with "the powers that be." If we succeed in controlling the masses, and preserve order at our meetings, we will not be likely to have any trouble "at court." But if, after all, we should be interfered with in the conscientious discharge of our duty, under the functions of our high commission, then return the apostolic answer, "Whether it be right in the sight of God to hearken unto you more than unto God, judge ye. For we cannot but speak the things which we have seen and heard."

VI. THE PREACHER IS NOT ABLE TO PREACH IN CHURCH AND OUT DOORS TOO, AND MUST GIVE THE PREFERENCE TO THE REGULAR SERVICES OF THE SANCTUARY.

Very well, if such is the fact in your case, I think you choose the right alternative. I would not advise you to neglect your regular appointments by any means; but yet, are there not very many who can, in addition to their regular appointments in church, take an extra one in the streets? I never have, nor do I now, sit in judgment on the consciences of my brethren in regard to this matter. Nearly the whole itinerant family are out-door preachers at camp-meetings and other extra occasions, and many of them preach themselves into a premature grave. I, nevertheless, believe that there are ten thousand ministers in the United States, among the different denominations, who are naturally well adapted to this work, and who, by proper application, would excel as street preachers, and fill their pulpits all the better for it. They have good voices for singing, a ready utterance, and a fair development of *tact* in the management of promiscuous audiences; and all that is necessary, is for them to feel that "Woe is me if I preach not the Gospel" in the streets, and go at it, and stick to it, till the Master says, "It is enough."

VII. It will give the preacher the bronchitis.

I give it, as my candid opinion, that your throat and lungs will suffer much less in the pure open air than they do in the carbonized, sickly atmosphere of crowded churches. I am accustomed to listen to the same clear voices in the streets, three hundred and sixty-five days in each year: "Fish! fish! fresh salmon!" "Eggs! eggs! fresh California eggs!" "Candy! Here's your celebrated cough candy! Everybody buys it; now's your chance!" "Here's your fresh California pears, apples, oranges, and peaches! Only two bits a pound! Buy 'em up!" "Latest news from the East! Arrival of the 'John L. Stephens!' Here's your New-York Herald, New-York Tribune, and New-Orleans True Delta!" Who ever heard of the fish, egg, candy, or fruit "crier," or the newsboys taking the bronchitis? An auctioneer will stand in the street, and "cry" at the top of his voice for two hours every day, and yet we never heard of an auctioneer taking the bronchitis. "He gets used to it." It is his business, and his physical functions adapt themselves to it. Rev. I. Owen and myself were, a few years ago, highly entertained for a few minutes, as we passed along tho streets of San Francisco, with the extraordinary earnestness of an auctioneer. I said to my friend, "If we could get ministers to 'cry aloud' as earnestly over living immortal souls as this man does over spoiled cheese at two cents per pound, what a waking up they would produce among the sleeping thousands of this land!"

VIII. It will crack and injure the voice.

If you will not bind your neck with a tight cravat, and if you will stand erect, head up, speak naturally, and not strain your voice, you will experience an improvement in the quality and an increasing compass and power of voice, and a greater facility in natural utterance by regular street preaching. Ten years ago, preaching two sermons in church and one in the streets, caused me hoarseness of voice and great weariness of body; but now, with three sermons in church and two in the streets, each Sabbath, I have no hoarseness, and but little weariness. Before I commenced street preaching, I was subject to violent colds and soreness of throat and lungs; but I have known, by experience, nothing of "sore throat" or "sore lungs" for years. I would not intimate that I am invulnerable to such affections; but I do believe that the danger is lessened, at least fifty per cent., by the out-door preaching.

CHAPTER IV.

SUGGESTIONS FOR A STREET PREACHER.

To my young brother who has made up his mind to "go out into the highways," and preach the Gospel, I would respectfully submit a few suggestions.

I. Read over your commission: "Go ye, therefore, and teach all nations, baptizing them in the name of the Father, and of the Son, and of the Holy Ghost; teaching them to observe all things whatsoever I have commanded you: and lo, I am with you always, even unto the end of the world. Amen." Then reassure your faith by a little Gospel logic, thus: 1. Am I an embassador of Christ? 2. Do I obey the orders of the Master, "teaching them to observe all things whatsoever he has commanded?" 3. The conclusion: "Lo, I am with you always, even unto the end of the world."

For what purpose is he with me? Is it not to speak through his unworthy embassador, to apply the word immediately to the hearts of the hearers, and to save NOW such as will come unto him? In your own mind and conduct lay these premises in the streets, and the conclusion will apply as logically and as certainly to preaching in the streets, as within consecrated walls. Let the argument be accompanied with an "unction of the Holy One," prompting you to say, "The love of Christ constraineth us" to "go out" to seek the lost, and preach to those who most need it. Then,

II. Act under the authority of your commission upon your own convictions of duty. Consult no man as to whether or not you should do your duty. You may inquire, if need be, where, in the streets, the greatest number of the "creatures" to whom you are sent, may be congregated, and what is the best hour in the day to get the best hearing; but to consult whether or not you should "go out," is, first, wrong in principle, because Jesus says *Go*, and thus fastens the obligation upon you, unless the condition of your health, or other providential bar, should operate to limit your obligation to preaching in the church; and, secondly, you will find in every place some excellent and pious men who will argue the inexpedience of street preaching in that place, and will thereby weaken your faith and purposes, and commit themselves to the negative of the question, against you. Whereas, if you simply announce your appointment for preaching in the street, and assign as the reason your convictions of duty in the premises, and bespeak the sympathy, and prayers, and coöperation of God's people, and invite all, saints and sinners, to attend, those who do not sympathize with the movement will not attend, but, not having committed themselves against it in advance, they will say but little, and give you no trouble.

III. THE "PREPARATION" FOR A SERMON IN THE STREET. You should have clear perceptions of the leading principles and facts you wish to announce. Let your propositions be briefly stated, in simple, appropriate language, and your principles be clearly defined. If you wish to employ arguments, let them be short, practical, and to the point. Illustrate the truth amply, and apply it promptly and pointedly as you proceed. Draw your illustrations from the every-day transactions and occurrences of life, as did the Saviour and his apostles. Make it a point, at all times, to gather up and store away suitable illustrations of Bible truth, from the streets, from the newspapers, hospitals, prisons, and from your pastoral visitations in domestic circles. Fresh facts, from personal observation, are much better in their effect than borrowed ones, or second-hand stories.

Do not confine yourself so closely to any system or arrangement of your sermon, as to prevent your seizing and laying under contribution all the incidents of the occasion which may serve to illustrate your subject. These spontaneous illustrations, seized impromptu, and skillfully applied, can hardly fail of a good effect upon the audience.

If you will bear with me, I will give you just here a few illustrations of this point. One Sunday afternoon in 1853, preaching on the "Long Wharf," and wishing to illustrate the distinction between a decent, well-behaved sinner, outwardly, and a violent, outbreaking sinner, I remarked, after stating the point, "Gentlemen, I stand on what I suppose to be a cask of brandy. Keep it tightly bunged and spiled, and it is entirely harmless, and answers some very good purposes; it even makes a very good pulpit. But draw that spile, and fifty men will lie down here, and drink up its spirit, and then wallow in the gutter, and before ten o'clock to-night will carry sorrow and desolation to the hearts of fifty families. So that man there, trying to urge his horse through the audience," all eyes turned from the cask to the man, "if he had kept his mouth shut, we might have supposed him a very decent fellow; but finding the street blocked up with this living mass of humanity, he drew the spile, and out gurgled the most profane oaths and curses. But, while there is now all the difference between outwardly moral and out-breaking sinners, as between a tightly-bunged and an open cask of brandy, I would invite your attention to a time when there will be no material difference between them.

"Should you attempt to get this harmless cask of brandy through the custom-house in Portland, Maine, the inspector would pay no regard to the outside appearance, or separate value of the cask: he would extract the bung, let down his phial, draw out and smell its contents; then shake his head, and mark it contraband. My friends, God has a great custom-house, through which every man has to pass for in spection, before he can be admitted into his kingdom. When you are entered for examination, do you imagine that the great omniscient Inspector will pay any regard to your outside appearance or conduct? Nay, my dear sirs, he will sound the inner depths of your souls. All who are 'filled with the spirit' of Christ will be passed, and treasured up as meet for the Master's use; but all who have not the love of God shed abroad in their hearts, will be pronounced 'contraband,' and branded eternally with, 'Depart, ye cursed, into everlasting fire, prepared for the devil and his angels.'"

On another occasion, near the same place, I was preaching on the bondage of sin, and said to the large audience assembled: "My dear sirs, you are slaves to sin and Satan; your conduct proves it, and frequently you unwittingly confess it. I said to a man a few days ago: 'My friend, you ought not to swear.' 'It's a free country,' said he, 'and I'll do as I please.' 'But, sir,' said I, 'a gentleman will not please to indulge in a practice so useless and wicked. Moreover, I don't allow a man to swear in the presence of my little boys here.' 'Well,' said he, 'I know it is a mean practice; but I've got into the habit of it, and I can't quit it.' So, in trying to apologize for your various sins, you have often confessed the fact that you are a poor prisoner in bondage to sin. A man enslaved by habits of intemperance came to see me a few days since, and said: 'Father Taylor, what shall I do? I have a dear wife and four sweet little children in New York, and I am afraid I shall never see them again,' crying as though his heart would break. 'I used to have plenty of everything I wanted, and was happy with my dear family; (God bless their dear souls, I fear I shall never see them again;) but I came to California, fell in with bad company, and have gotten into this cursed habit of drinking, and can't quit it. I've tried often; but it's no use.' 'Now, my friend, said I, 'for the sake of your family, that you say you love, for the sake of your poor body, so much abused by rum, and for the sake of your soul, redeemed by the blood of Jesus, do make one more effort to be a man. Shun your drinking companions as you would Satan, and fly from the grog-shops as you would from the yawning mouth of hell; and cry to God in the name of Jesus for pardon and help.' 'I will, Father Taylor, I will. So help me God, I'll never drink another drop.' The very next week I found him drunk in the streets.

"One such, came to me a short time ago; and after relating the sad tale of his sorrows, asked to sign the pledge. I gave him a pledge, and he signed it, saying: 'There it is; my name is there once for all. Henceforth I'll be a sober man.' The next day, as I passed up California-street, I saw him with a demijohn in his hand. 'Why, my friend,' said I, 'what are you doing with that stuff?' 'O,' said he, 'I thought, as I was knocking off for good this time, I would just take one more nip.' My dear friends, such is your own bondage to your prevailing sins, whatever they may be. Chains of habit are stronger than chains of steel; you cannot break them."

Just at that moment, a candidate for the chain-gang was conducted along the street, with a heavy chain around his leg. Said I:

"Look at that poor fellow! How gladly would he kick off that heavy chain, and be free! But look at that great band of iron round his leg, and the strong links. He cannot break them. And yet he is no more a prisoner to-day, under that heavy chain, in the hands of his keeper, than you are under the chains of sinful habit, in the hands of your keeper, the devil, by whom you are 'led captive at his will.' 'O, well,' says one, 'if that be true, it is no use for us to try to be better, and you had as well let us alone.' That such is your bondage to sin, there is not a question, your own consciences and the word of God being judges; and your utter inability to free yourselves is equally true. You may, to be sure, under certain helpful influences, break off from some of the outward forms of sin, but not from sin itself. You have tried it often, and failed every time. 'What, then, shall we do?' says one. Ah! I have you now just where I wanted to get you; where the Philippian jailer was when he cried to Paul and Silas, 'Sirs, what must I do to be saved?' Into the self-conscious bondage which St. Paul describes in the seventh chapter of his Epistle to the Romans, 'I consent unto the law that it is good.' 'But I see another law in my members warring against the law of my mind, and bringing me into captivity to the law of sin, which is in my members.' The law of sin in your members, sinful propensities, passions, and habits. Do you understand the practical workings of that law? 'O, wretched man that I am, who shall deliver me from the body of this death?' Not only imprisoned, but bound to a dead body, face to face, and limb to limb. Who shall deliver me from such bondage and death? Thanks be unto God, there is an almighty Deliverer now waiting to be gracious. 'God, through Jesus Christ our Lord,' can, and will now deliver you, if you call upon him while he is near."

On the occasion last referred to, the wind being high, there was a sudden cracking noise heard among the shipping, and a part of the audience started to run and see what the matter was, when I said:

"What a dreadful thing it would be for some old ship to be wrecked. You would talk about it all day, and to-morrow morning all the papers would herald the sad disaster and loss; but souls, precious souls, one of which is worth more than all the fleets and navies of the world, are wrecked in our midst dayly, and drift down the gulf stream of despair into the maelstrom of hell, and nothing is said about it, no paper announces the sad disaster; a soul wrecked and damned forever! no possibility of recovery, and no insurance."

IV. In regard to the management of an out-door audience, I would remark:

1. If you apprehend disturbance, put every man on his good behavior as an American citizen, or as persons who have had some advantages in "good breeding," and who have self-respect, and presume that good order is what you expect, as a matter of course. If a man misbehave, always speak kindly to him. Appeal to his reason and common sense, and

if he has any soul in him, and not too much rum, you can do anything with him you please. Some fools have to be answered "according to their folly." One Sunday morning, as I was preaching on Davis-street, a fellow came close to the barrel on which I stood, and looking up into my face, said:

"The apostle David says, it is hard for thee to kick against the pricks."

"See here, my friend," said I, "when did you arrive, sir?"

"I came from the old country," said he, "about six years ago."

"But I want to know when you came to California?"

"O, a good while ago," said he.

"How many days since?" said I. He hesitated, and looked for an opening through the crowd, by which he might escape, and then replied:

"About two weeks ago, sir."

"I knew," said I, "by your conduct that you had recently arrived, and had not learned how to behave yourself here yet. You seem to imagine that we were all a set of heathen here in California, and that you could 'cut up,' and do as you please. Now as you are a stranger in these parts, I will inform you that the order of the day in California is for all classes of society to respect the preaching of the Gospel, and never to disturb a preacher in the discharge of his duty, and the fellow that dare persist in such an outrage may expect that even the gamblers will 'give him a licking.'"

I have often caused men, when trying to make a disturbance, to run and hide themselves, by offering an apology for their conduct. "Don't hurt that poor fellow, friends; we must make great allowance for his bad conduct. It is fair to presume that he has just arrived from some barbarous island of the Pacific, and has not yet learned how to behave himself." To turn the eyes of an audience, sparkling with good-humored contempt, upon a fellow, will move him as suddenly, almost, as a charge of bayonets. I have, however, always run such fellows off the track so good-humoredly, that I have never yet had an after difficulty with one of them.

2. If by a cry of fire, or otherwise, your congregation is scattered, do not be discouraged, but watch your opportunity to take advantage of the disturbing excitement, and set your sails to take the breeze· and you will, probably, double or quadruple your congregation in five minutes, and then, under the excitement of the occasion, thunder home the truth into the wakeful, curious minds of the crowd. An important point is gained, when, by any legitimate means, the people are fairly waked up, so as to listen attentively. Get your metal melted, and then mold it. I might produce a hundred illustrations

of this subject from real life; but a few may suffice here, as they will be interspersed through the subsequent pages of this work. When preaching in Georgetown Market, in 1846, on one occasion, a sudden noise was heard up the canal, near where we stood; and it was rumored that a boy was drowning. As the congregation ran, I sang; and in one minute they were all back, and quietly waiting for the remaining part of the sermon. Once, in Belair Market, Baltimore City, in 1848, I was about half through my discourse, when a large funeral procession passed by, accompanied by a "band of music." The melody of the "band" took the ears of my audience; and as I saw them beginning to break away to join the procession, I said, "Brethren, we can make better music than that;" and struck up the best song at command, in which the "congregation" heartily joined. O the pathos and melody of that song! We heard no more of the "band of music." The result was that I held the audience; and a friend standing out where he had a good view, said, "At least one hundred of the procession broke rank, and came to the preaching."

In 1854, on the Plaza, in San Francisco, just as I was reading my text, a Frenchman stole a pair of boots on the opposite side of the Plaza; and the cry rang through the streets, "Stop thief! stop thief! stop thief!" causing one general rush. Seeing that I could not directly withstand the force of the tide, I said, "Run, boys, run and catch him! Put him into the station-house, and hasten back. I've got something to tell you. I'll sing again, and wait for you." By the time the song was ended, back came the crowd, doubled, and multiplied by the addition of all the thief-catchers within a dozen of squares. I then, as I always do, tried to improve the occasion, saying, "Gentlemen, the devil helped that poor Frenchman into a bad job, when he stole those boots. The old fellow is very sharp, and does not always design to get his servants into such troubles; for he wishes to tie them permanently to his interest, and lead them quietly down to hell. While you look with contempt upon the poor boot-stealer, you forget that many of you are equally dishonest, only you steal in some more honorable way. And you overlook the fact that most of you are guilty of the outrageous crime of 'robbing God.' The devil tries to blind you to that fact, until you exhaust the patience of God, and fill up the measure of your iniquity, and then, when the righteous God delivers you over to your master, whose companionship and service you have chosen, the same smooth, diabolic tongue which deceived our first parents, and now lures you along so charmingly in the way to hell, will then, in tones of thunder, pursue your frightened souls through the caverns of dark damnation, and ring the cry eternally in your ears, 'Stop thief! stop thief! Catch him! catch him?'"

A cry of fire has often elicited appeals of this character: "Why, my friends, the devouring fire is a dreadful thing. To see the labors of years consumed in an hour, and poor families turned out homeless and friendless. But O, my God! what are all the disasters of fire, here, compared with the interminable fires of hell, which will soon break out upon the souls of most of my audience, unless they fly to Christ for refuge. Who among us 'shall dwell with the devouring fire?' who among us shall 'dwell with everlasting burnings?'" Isa. xxxiii, 14.

Once on the Plaza the congregation was disturbed by a false alarm of fire, and I said: "My dear sirs, how quickly a cry of fire, though often, as in this case, a false alarm, starts you. You run as though the salvation of the world depended on the race. I come to you here every Sabbath with an alarming cry, the danger of which, I warn you, is more dreadful than the burning of all the cities on the globe at one time, and I never raise a false alarm. I cry, Fire! fire! fire! hell fire! It is breaking out in our very midst every day, and sweeping down the souls of your neighbors into the hopeless depths of the burning lake beneath, whence 'the smoke of their torments will ascend forever and ever!' Why do you not run, and fly as from the brink of hell, and take refuge in the

-left Rock of the Gospel, the 'Rock of Ages,' in which if you abide, the conflagration that shall consume the universe shall not singe a hair of your heads? The decisions of this day may probably decide the question with you forever."

Attending the Petaluma Camp-meeting a short time since, while a brother was preaching very earnestly, a horse broke loose in the rear of the preacher's stand, and making a great noise among the wagons, the people sprang to their feet, *en masse*, and many started to run. The preacher stood confused for a moment, when a loud voice from the stand said: "What a dreadful thing it would be for an old horse to run off and break his neck, but for a few immortal souls to go down to hell is a very small matter, brethren. Go ahead with your sermon, brother." The people bent to their seats almost as suddenly as if they had been shot at, and the preacher proceeded with his discourse.

In the summer of 1855, I had an appointment to preach one week-night, in a large bar-room on Moor's Flat, in the mountains. The congregation assembled early, and spent an hour in playing ball. When the bell rang for preaching the mass of the audience assembled on the porch, and "cracked jokes," and sang lewd songs, with the design, I thought, of intimidating the preacher. After letting them conduct the exercises in that way for a few minutes, I said: "Hold on, boys, and let me sing you a song." They gave audience, and I did my best on one of my best pieces. Nothing could be more calm than the salubrious atmosphere on that occasion, and the surrounding mountain heights, and deep "canyons," and giant trees of the dense forest, all combined to render the scene impressively grand and solemn. The echoes of the song came back from the neighboring mountains, and the trees seemed vocally to be praising God in song. The singing ended, I said: "Now, boys, walk in here, I have something to tell you." They all slipped in as quietly as possible, and I had a blessed season in pressing home upon their hearts the word of life.

In July, 1855, I spent a Sabbath in New Orleans, a beautiful mining town, high up in the mountains of California. It was said that a copy of the anti-gambling law, which had been passed at the late session of the Legislature, had not been forwarded to the authorities in that place, and therefore did not take effect in New Orleans, in consequence of which it was said that nearly all the gamblers of those mountains had assembled in that town to carry on their business. During my short stay with them I preached four times in the streets, and once in a private house. They listened to me in the street three times with marked respect and attention, but when, on Sunday afternoon, I took the street, and commenced to sing them up for a fourth hearing, they seemed to think that they had had "enough of a good thing," and that they would "run me off the track." So they got up a boy-fight near by in the street, between an American boy and a Spaniard, and the cries rang, "Huzzah for Young America!" I sat down on my "goods-box" pulpit, and waited till the fight was over, and then arose and commenced to sing again. The fight had attracted a dense crowd, and the thing I had to do, was to take them in the name of the Lord, and "compel them," if not to "come in," at least to listen attentively to the invitation sent out by the Master.

As I was engaged in drawing the crowd with the second song, the fellows next "got up" a dog-fight, and at it they went hissing and whooping, when I said, "Run, boys, run! We are all seeking enjoyment, and trying to be happy! There's a rare opportunity! You are under a high excitement of animal feeling! A glorious entertainment that! What an intellectual feast it must be to enlightened, high-minded American gentlemen, to see a couple of dogs fight!" By that time I had the last man of them, and the good-natured dogs, having nobody to prompt them, concluded not to fight, and trotted away together; but their prompters all remained to listen, and I proceeded, saying: "Gentlemen, I do not blame you for seeking enjoyment, and for trying to be happy. God, who made us, and endowed us with wonderful powers of intellect and heart, designed us to be happy, and hence this insatiable thirst for happiness which constitutes the mainspring of human action. The difference between us is in regard to the source whence we may derive substantial happiness, whence the demands of these quenchless longings of our souls may be met. You have tried a great many sources, money making and money spending, rum drinking and gambling, with occasional boy and dog fights. Bills were posted all through your streets last week, promising a rich treat for immortal souls, on the Fourth of July, in American Valley. The intellectual feast to commence with a fight between a bull and a grizzly bear. The second course to consist of a 'magnificent dinner,' and as much whisky as could be desired at two bits a 'nip.' The third course to consist of music and dancing among the men, (ladies were very scarce,) which might be protracted till every soul was satisfied. Your undying spirits were so hungry and restless that you could not let such an opportunity pass, so away you went to American Valley. To your great disappointment, the bull and bear had determined to remain friends, and would not fight. The dinner was good, the whisky was very bad, but you thought you would make it up in the ball room, so you kicked round there for a few hours, and, stopping to rest your poor bodies, you looked within to see if your souls were happy. Poor souls!

they were disappointed, and faint with hunger, and you said to yourselves, 'Well, there must be something in it; these other fellows seem to be happy, so I'll try it again.' At it again you went, and shuffled round there till the dawn of the morning, and the next day your pockets were minus $100 each, bodies worn out with exhaustion from want of sleep, excess, and riot, and your souls, what shall I say of them? A more miserable set of fellows can hardly be scared up this side of perdition! So much for your pleasure taking and intellectual felicity! Now the repetition of these things, through succeeding years, with invariably the same miserable results, ought to convince you that you are on the wrong track, and that, continuing the same course, your souls will continue to be the dupes of disappointment and remorse all through your probation of life, and then have an eternity in hell for the hopeless repentance of your folly." I then, in my sermon for the occasion, proved to them that God alone, through the mediation of Jesus Christ, was the source of substantial comfort for immortal souls, and that nothing but experimental religion could make us really happy in this life, or in the world to come.

I believe that God's Holy Spirit applied the truth, and touched many of their hearts, for some of them wept like children, and all listened with great apparent interest. I ask pardon for giving here more cases of illustration than I intended when I set out They pressed themselves upon me, and I have admitted them because of their variety as to time, place, and character.

In conclusion, I will add that, after all, you should make up your mind, as a street preacher, to be considered and called "a fool" for Christ's sake, and to be grinned at by the scorner, gazed at by the multitude, "sighted" at by gentlemen through hand-glasses, double-barreled spy-glasses, and large telescopes; to be sworn at by ruffians, and to be slandered by many you call your friends. But never mind, trust in God, and do your duty. Rely for success alone, both for the use of means and the attainment of desirable ends, upon the merit and intercessions of Jesus Christ, and the Divine efficiency of the Holy Spirit, and you will praise God through eternal ages that you were, by his grace, enabled to "preach the Gospel to the poor" in "the streets and lanes of the city," and in "the highways and hedges."

INSIDE VIEW OF THE GOLDEN GATE

CHAPTER V.

PRIMITIVE CLASS-MEETINGS IN SAN FRANCISCO.

In the early days of Methodism in this city, I had a general class-meeting in the chapel every Sunday afternoon, at which there were usually present from fifty to ninety persons. There was then but one "charge" in the city: no "North," no "South," no party differences nor jealousies of any kind. There was a constant stream of emigration flowing in through our "golden gate" from every part of the world.

The city was small, so that the "royal proclamation," sounding out from the Plaza every Sunday, tapped the drum of nearly every man's ears in town. All the Methodist passengers, and multitudes besides, immediately showed their faces. After proclaiming to them a crucified and risen Jesus, I always announced the appointments for preaching and class-meeting in our "church on the hill." Hence the size and variety of our class-meetings. As a specimen, I extract in substance the following notice from my journal, dated Sunday, February 3d, 1850:

"There were in class to-day about ninety persons, witnesses for Jesus from almost all parts of the United States, from Maine to Texas; and from Buenos Ayres in South America; from Costa Rica in Central America; from Prince Edward's Island; from England, Scotland, and Ireland; from Germany, Sweden, and Denmark; from North Wales, New South Wales, and New Zealand. They all uttered distinctly the shibboleth of Methodism and told the same story of 'redemption through the blood of Jesus, even the forgiveness of their sins.'

"A very common inquiry in the mouths of Wesleyan Methodists from England and her colonies was, 'Do you belong to the Church that Mr. Wesley established in America—the Church of Mr. Asbury and Dr. Coke?' So soon as they heard the answer, 'Yes,' they immediately extended the 'right hand of fellowship' for another greeting, and, with tearful smiles, uttered with great emotion, 'God bless you. It is quite an unexpected pleasure to meet you here.' An observing stranger, beholding the scene, would have said, 'No doubt there is a meeting of two brothers, sons of the same mother, who have not seen each other for twenty years.' And brothers we were with a free good-will, bound together by bonds of mutual sympathy and Christian affection stronger than ties of blood, though we had never seen each other before, and probably never would again till the great reunion of the blood-washed brotherhood on the other side of the river."

At the class-meeting above referred to, an old gentleman, with a long, gray beard, by the name of Livesey, (I do not remember that I learned his Christian name,) arose and shouted the praise of Jesus, and thanked God for full salvation "through the blood of the Lamb." He thanked God, also, for Methodist class-meetings, which, for thirty years, had always been seasons of refreshing to his soul. Thirty years ago from that day he had obtained the forgiveness of his sins, and had never turned his back on Jesus. Heard Dr. Adam Clarke preach a sermon on "*Hope*," "Which hope he had as an anchor of the soul, both sure and steadfast." Had always been a firm believer in the doctrine of holiness as taught by Mr. Wesley, and yet, continued he, "strange as it may appear, I never obtained an evidence that I was *wholly* sanctified till last Tuesday night. I was aboard ship in the harbor out there, and while all hands were locked in sleep, and nothing was heard but the dash of the waves against the sides of our vessel, my soul was waiting upon God, in an unusual exercise of prayer and faith in Christ, when the power of the Holy Spirit came upon me as I never felt it before. I realized an application of the all-cleansing blood of Jesus to my heart, and that I was made clean through the Word. My soul has been full of glory ever since. We have pitched a tent on the beach in 'Happy Valley' for prayer-meetings, and God is with us there. Glory! glory be ascribed to his holy name!"

The old man took his seat with subdued utterances of "Glory! glory! glory be to God!"

After that meeting I saw his face no more. During that week he left the city on business, and word came back that his vessel was capsized in the San Joakin River, and that the good old brother was drowned. Never learning anything to the contrary, and receiving additional confirmatory evidence of the truth of the rumor, I settled on the conclusion that God, who buried the body of Moses in some unknown spot "over against Beth-peor," had deposited the body of Father Livesey in some one of the mighty eddies of the San Joakin River "until the redemption of the purchased possession." His spirit, we doubt not, has gone to bathe in that "pure river of water of life, clear as crystal, proceeding out of the throne of God and the Lamb."

Although I had but a very limited acquaintance with Father Livesey, his image is very distinctly defined in my memory, and I believe I shall recognize him on the other side of Jordan, when, through the great mercy of God, I shall have reached that shore, and shall hear from his own lips the mysterious manner in which God in his wisdom took him from labor to reward.

At the class-meeting in question many thrilling experiences were related. At least six persons bore a clear testimony to the all-cleansing efficacy of the blood of Jesus applied to their own hearts.

CHAPTER VI.

MY FIRST PREACHING IN THE STREETS OF SACRAMENTO CITY.

I SPENT the first Sabbath of October, 1850, in Sacramento City, and had the privilege of preaching three times in our "Baltimore California Chapel," so called because our kind Baltimore friends framed it, and paid for it, and sent it to California. It was destroyed by fire about three years since, but the Sunday school organized in it in 1849, now (July, 1856) numbers upward of three hundred scholars. In the afternoon of the said Sabbath day, I selected a goods-box on the "levee" for a pulpit, and opened my commission for the first time in the streets of that city. While singing the "royal proclamation," two men rode up near to where I stood. I never learned their names, but, for convenience, will call them Bacchus and Fairplay. Bacchus was pretty drunk, and began to yell and make a great ado. Judge W. and a few others took hold of his mule's bridle, and tried to lead him away.

"Let me alone," cried Bacchus.

"Let go his bridle," said Fairplay. "This is a public street, and you have no business to interfere with him. Let him go, I tell you. If you don't let him go I'll see that you pay dearly for it." And many other hard threats were uttered by Mr. Fairplay.

The singing, which had been continued without interruption, together with the strife and hallooing of the drunken man, attracted an immense crowd. When the opening hymn was ended, Judge W. and his companion had gotten Bacchus off to the distance of about thirty yards, and had about equally divided the crowd. At that moment I called to the judge and his company, saying: "If you please, gentlemen, let him go, and I'll take care of him." But they had become so zealous in the matter that they seemed determined to drag him away, and would not let him go. By the time I had sung another song of Zion, they had gone but a few feet further off, and had half the audience, who appeared to be more interested in the fate of the drunken man than in the songs of the preacher. I then called to them again, and said: "Gentlemen, you had better take my advice. If you will let that man go, I will send him away in one minute. I am surprised at you Sacramento folks. Come down to San Francisco, and attend preaching on the Plaza next Sunday afternoon at three o'clock, and I'll show you how to behave. Men naturally

5

run after an excited crowd, but you have all seen the great attraction, a drunken man on a mule. Now, let me manage that fellow, and all of you come up here; I've got something to tell you."

With that they let Bacchus's mule go. I then addressed his threatening, storming companion, Fairplay, and said: "I deliver that man up to you, sir; I want you to take charge of him, and lead him away. Take good care of him, if you please."

"Yes, sir," said he, "I will," tipping his hat as he made his best bow, and immediately led him away. The whole crowd then gathered round me, and I said, "Gentlemen, some of my friends here say that it is getting too late for preaching this afternoon; that by the time I get under way the supper 'gongs' and bells will ring, and that you will all run off to supper. I have some very important things to say to you, and I will have done before the tea gets cold. Now you had better stay and hear me out, and my friends here will find that they are not so good at guessing as they thought they were."

I then announced as my text, "Godliness is profitable unto all things, having promise of the life that now is, and of that which is to come." The preliminary exercises seemed to have raised the temper of their minds to an impressible state, and the power of God's Holy Spirit manifestly attended the word. Many eyes unused to weeping, gave forth their briny streams. Good order and great solemnity pervaded the entire assembly. The supper gongs in the neighborhood set up a prodigious ringing before I had got half through, but I saw none leave. All seemed willing to risk the "cold tea." After singing the Doxology, all hats off, many strangers gathered round me, and wept as they told of their sorrows, and inquired about Jesus, the sinner's friend.

CHAPTER VII.

STARVATION IN AN EX-CITY HOSPITAL.

In the month of February, 1850, the "City Council" made a contract with Dr. S., by which all the sick of what had been known as the "City Hospital," kept by Dr. M., should be removed to a new hospital, fitted up by Dr. S., on the corner of Clay and Powel streets. Dr. M. said that the term of his contract with the city for the care of her sick had not expired, and that he would not give up the patients, and that he would institute suit against the city for breach of contract and payment for the care he should subsequently give to all the sick he could retain in his hospital. The "alcalde"—before we had a mayor—sent an order to Dr. M.'s hospital, requiring the sick to leave and repair to Dr. S.'s hospital, otherwise they would get no support or care from the city. Dr. M. told them to stay with him, and he would see that the city should support them, otherwise he would support them himself. More than half the patients obeyed the order; the rest remained in care of Dr. M. As the prospects of Dr.

M.'s suit against the city declined, his patients who were able to walk left him, and were admitted on new "certificates" into Dr. S.'s hospital. On the sixth of March I learned that Dr. M. had, on that day, notified his patients that he could keep them no longer. Whether from doubt as to the success of his suit, or want of funds, or the hope that the city authorities would have the patients removed, and by thus recognizing them as city patients, strengthen his plea for recovering his claims for keeping them, I know not; but he ordered that no more food or medicine be given them. The number of his patients had at this time been reduced to thirty men. Those who were able hobbled out, but there were left still seventeen men who were scarcely able to raise their heads from their pillows. The names of these were Franklin Baxter, James F. Dixon, John Raffsay, Charles Johns, Richard Johnson, Guslaff Myers, Samuel Howard, Charles Johnson, Joseph M. Gustin, C. C. Kindred, William Orr, Thomas M'Donald, Thomas Crosby, W. H. Reed, James Thompson, John Dixon, and a Frenchman whose name I did not get.

That night we had a meeting of the Executive Committee of the "Strangers' Friend Society," which had been organized in our chapel on the fifteenth of January, 1850, and in which all the Protestant Churches in the city were represented, and through which many sufferers were relieved during that very severe winter. At this meeting we appropriated the necessary funds for the relief of the said sufferers till they could be removed to the other hospital. J. B. Bond, son of the late Dr. Thomas E. Bond, of precious memory, was appointed a committee to wait on the alcalde, and make arrangements for their immediate removal. We had with us then an old gentleman by the name of Alfred Roberts, who came to California for the avowed purpose of devoting all his time to waiting on the sick gratuitously, and who spent his time accordingly. He was entirely without money himself, and yet made many, if not rich, at least comfortable in their destitution. He would accept no reward for his services, and yet all his wants were supplied. The said funds of the Strangers' Friend Society were placed in the hands of Brother Roberts, with orders to go the next morning, and supply the patients with whatever they needed, till they could be taken to Dr. S.'s hospital. About two o'clock in the afternoon of the next day, namely, the seventh of March, I went to see Brother Bond, to ascertain how he was succeeding with the alcalde, and learned from him that his efforts had not been successful; that while the alcalde had no objection to their going to the new hospital, he would not recognize their present relation as city patients, for the reason that Dr. M. would take advantage of that recognition in the prosecution of his suit. Brother Bond advised me to get a physician to examine the patients, and give me a new "certificate" for each one; and he thought I might prevail on the alcalde to give them "permits" to Dr. S.'s hospital, on the new certificates, without recognizing them as city patients.

I employed Dr. Hill, and went to examine the sick men, and there learned, to my astonishment, that they had had no food or medicine for twenty-four hours; that Dr. M. had forbidden Brother Roberts to bring anything into the hospital for the patients. His object seemed to be, to force the alcalde to remove them from his hospital, and thus tacitly acknowledge them as city patients. He remarked to Roberts that "if he allowed them to be fed and cared for in his hospital, he never would get rid of them."

I presented Dr. Hill's certificates to the alcalde, and pleaded for the poor fellows' lives, but he said he would have nothing to do with them while they remained in that hospital. By this time it was nearly night. House room was very scarce in the city in those days, and bedding for so many men could hardly be found anywhere, except in the hospitals and hotels. Hotel keepers would have nothing to do with hospital patients. The poor sick men, worn down by disease and hunger, seemed to forebode the worst that could befall them. Old Captain Baxter had braved the thunder of a thousand storms, but now his courage failed. Said he, "I give up. It's no use; it's no use. I can stand it no longer."

William Orr was an Englishman, from the West Indies. He had large, keen black eyes, and silver locks of hair, and looked as venerable as a bishop. During his many weary weeks in the hospital, he had frequently sent for an old acquaintance, a wealthy man from the West Indies; but his "familiar friend, in whom he trusted," would not see him. Mr. Orr said to me, concerning his friend, "When G.'s rich uncle in C., who had an estate to be inherited, was in his last illness, G. never left his bedside, day or night; and if I had property, as I once had, he would come very quickly; but he is afraid I might ask him for assistance." The old gentleman was generally calm and self-possessed, but this strait seemed too much for his feelings, and he said, "Well, have I come to this at last? Could I ever have believed it? O, Christ, have pity upon me in my low estate!"

Near him lay Thomas Crosby. He had the dropsy; his body was as big as a barrel, and he had occasional spells of suffocation, from which it seemed impossible for him to rally; but he was triumphant, and spent his time in laughing, and weeping, and praising God. "O what a precious Saviour I have found," he would say. "Glory be to Jesus! I shall soon be done with pain and sorrow! I shall see Jesus!" I believe he was nominally a Roman Catholic, but he had received a holy anointing that did for him what an ocean of holy water could not do.

I mention these cases as specimens of the scene. But the question was: "What shall be done to relieve these sufferers?" I obtained from Dr. M. the loan of the cots and beds they occupied for one night. I rented a house in the neighborhood, and hiring some help, we took up the beds with the patients on them, "and walked" to our "hired house," where our friend Roberts provided everything possible for their comfort, and took care of them through the night. The next day, by a little maneuvering with the alcalde, to dodge a technicality, I succeeded in getting them all into Dr. S.'s hospital, where most of them afterward died.

C. C. Kindred, and I believe two or three others, recovered. Captain Baxter often spoke of his pious wife, "as good an old woman as ever lived," and how for many years she had been praying for him. He only wished that he were as good as she. He was sincerely penitent.

James F. Dixon was from Louisiana. He professed religion several weeks before his death, and seemed to be fully ready. Crosby went home shouting. G. M. left the world swearing. Joseph M. Gustin was happy in God, and had a peaceful hour in which to die.

I took great pains to point each of them to "the Lamb of God that taketh away the sin of the world," and hope that others of them found their way to that healthful clime, where the people never say, "We are sick." "There God shall wipe away all tears from their eyes; there shall be no more death; neither sorrow nor crying; neither shall there be any more pain; for the former things," among which are those dark hospital scenes, shall have "passed away."

CHAPTER VIII.

A BROADSIDE UPON THE ARMY OF THE ALIENS.

On Sunday afternoon, the 5th of May, 1850, I took my stand upon the porch of the "Old Adobe," on the Plaza, and after singing up a crowd of about a thousand persons, I announced as my text, the fourth and fifth verses of the one hundred and fortieth Psalm. "Keep me, O Lord, from the hands of the wicked; preserve me from the violent man; who have purposed to overthrow my goings. The proud have hid a snare for me, and cords; they have spread a net by the wayside; they have set gins for me."

Before me lay a vast scene of desolation; for on the day preceding, at four o'clock in the morning, the dwellers of our city were aroused from their slumbers by the cry, "Fire! fire! fire!" It commenced in the United States House, on the east side of the Plaza, within a few feet of where a fire broke out in December, 1849, in the midst of the "gambling hells."

For an hour or more, nearly everybody seemed to stand back aghast, and silently watch the devouring element, as it swept block after block of the best buildings in the city. Three entire squares, with the exception of three or four houses, were consumed within the space of four hours. The loss was variously estimated at from three to five millions of dollars.

In the elucidation and application of the text announced, the Lord assisted me greatly in exposing the snares, and pits, and gins, (gambling-houses, grog-shops, and houses of prostitution,) and the wicked "and violent" men employed, with all their deceitful, attractive appliances and "cords."

While special thunder was thus being dealt out, a man on horseback gathered a crowd on the opposite side of the Plaza, and marched up, as though he intended to make a charge upon us. But the truth, peal after peal, continued to mount the wings of the wind, and make the sinners quake in its onward flight, so that our opposing general, by the time he reached the outer circle of our crowd, was awe-struck, and beat a quick retreat, leaving his men in our hands, who remained quiet and orderly listeners till they were regularly dismissed. We warned the people to beware of those snares, and pits, and wicked men, and urged them, as their only sure means of safety, to adopt the Psalmist's prayer, "Keep me, O Lord, from the hands of the wicked; preserve me from the violent man," etc. The power of the Lord was graciously manifest on the occasion. The day of eternity will exhibit the fruit.

CHAPTER IX.

THE IRISH SAILOR'S DILEMMA.

On Sunday night, the tenth of March, 1850, at the close of sermon in our "Little Church on the Hill," an Irish sailor came to the altar, in presence of the congregation, and said he wanted a word with the captain, meaning the preacher. I shook his hand, and asked him what I could do for him.

Said he: "I want you to teach me. My mother was a poor widow, as are the mothers of a large proportion of sailors, and she didn't know what to do with me. She couldn't take care of me, nor teach me, so she sent me to sea when I was a little boy. I have been to sea ever since. I am now thirty years old, and have niver had ony teaching. Now, I want your riverence to teach me."

"You have learned to use strong drink occasionally, have you not?"

"O, yes; I takes a wee drap sometimes."

"And you've learned to swear too, I suppose?"

"O, yes, sir, I've been a very bad man; but now I wants you to teach me how to be a good man."

I then explained to him his wretched condition as a sinner, and gave him a few lessons in "the first principles of the oracles of God," and urged him to fall out with his sins, and renounce them forever, and accept of mercy through a crucified Saviour.

"I thank your riverence for your good advice. I'll try from this very hour, and do as you say."

He turned away and went immediately out of the church; but within two or three minutes he returned and said:

"Your riverence, you'll pardon me, but I've thought of another thing I want you to tell me about. I've heard that the Bible says, if a man strike ye on one cheek, you must turn round and let him strike the other! Now, does it say so?"

"That is the doctrine," I replied, "that Jesus taught his disciples; but that is a hard lesson for you to learn now. If you will practice the lessons I have given you, and pray to God in the name of Jesus Christ until you obtain the pardon of all your sins, you will love God so much for his great mercy to you, that you will not feel like fighting an enemy. You will feel that as God has forgiven you so many thousands of sins, you, too, can forgive those who trespass against you. And then, you will be so anxious to have everybody get acquainted with Jesus, that you will want to pray for your enemies, that they may find pardon too."

"But," said he, "if a man knocks me down to-night on my way home, what must I do?"

"You need not distress yourself," I replied; "if you go along and attend to your own business, nobody will trouble you. And if you earnestly seek God, as you have promised to do, he will take care of you, and will not let such a hard trial come upon you at the start."

"But faith, and maybe he might strike me to-night on my chake, (cheek;) then I must turn the other chake, and let him bang away at that, ah? That's hard."

I tried to impress on his mind the importance of learning one thing at a time, and not to perplex his mind with the hard lessons at the last of the book till he had learned all before them. I presume he "shipped" the next day. I have not seen him since.

How many, alas! who have enjoyed all the advantages of a Christian education, and who have even "tasted the good word of God, and the powers of the world to come," fall into the Irish sailor's dilemma, and take the "wrong horn," moreover, by returning evil for evil.

CHAPTER X.

"THEY'LL THINK I'M A THIEF."

In the fall of 1850, an intelligent Swedish sailor was one Sunday afternoon attracted to the Plaza by one of the street preacher's songs. The words of truth, which fell on his ear on that occasion, so affected his heart, that he determined to follow the preacher, and see where he dwelt. He did not then make himself known to the preacher, but followed close after him till he saw his residence, and the church in which he preached. That night he attended preaching in that church. After sermon an invitation was given, as usual, for all persons who desired to seek salvation and become acquainted with Jesus, to come forward to the altar. A soldier, late from the Mexican war, who had been awakened on the Plaza, immediately presented himself at the altar as a penitent. He was converted to God, and afterward became a zealous Christian and a local minister.

When our sailor saw the soldier go forward, he said to himself, "O, I wish I could go there too, but if I go there they'll think I am a thief, and I never stole anything in my life." The adversary of souls took advantage of a peculiarity in his education, and thus kept him back.

In Sweden all are taught to read and write, and at a certain age, upon a repetition of the Catechism from memory, and satisfactory examination as to character, are all admitted into the Established (Lutheran) Church. Thus the whole nation receives the elements of an intellectual and moral education, and becomes, at least nominally, Christian. Our nation might draw some useful practical lessons on education from old Sweden, especially as to the moral training of all her children. But when a Swede becomes guilty of theft, the penalty of the law is first executed upon him; he is then conducted to church, and is placed by the minister in a conspicuous part of the house, where he is questioned by the minister, who, after receiving promises of reformation, calls the attention of the congregation to him, and bespeaks their forgiveness and sympathy, and asks an interest in their prayers on behalf of the poor thief. Our sailor friend had seen this operation in his fatherland, and thinking that the same rule applied everywhere, concluded that if he went to a Methodist altar, it would expose him to the unjust charge of theft. So he went not. The following Tuesday night he attended class-meeting at "our house on Jackson-street." Near the close of the meeting he arose and said: "My dear friends, I am blind; I cannot see. O, the horrors of this darkness that settles down on my soul! I feel that I am a dreadful sinner, and I am afraid there is no mercy for me! You who are near to Jesus, please speak to him for me. I'm so far away, he won't hear me, but if you will speak to him for me, he may hear you, and have pity on me." So we all kneeled down and spoke to Jesus for him. Two days afterward he found Jesus, and became a sincere disciple. We have often been much delighted and edified by his experience in Divine things, and original expositions of Scripture, an example of which I here submit. One night, at a meeting convened by a regular weekly appointment to talk and pray together on the subject of holiness, our Swedish brother arose in his place, and said: "My brethren, I was just thinking of what Jesus said to Peter: 'Simon, Simon, behold, Satan hath desired to have you, that he may sift you as wheat: but I have prayed for thee, that thy faith fail not.' The Saviour does not tell him that he should not be sifted, but gives him to understand that he should be sifted, but adds: 'I have prayed for thee, that thy faith fail not.' Satan hath desired to sift us also, and we may all expect to be sifted, but Jesus hath prayed for us. The exercise of faith in him brings holiness, and holiness expands the soul, so that when we are sifted, we will not fall through the sieve. Without faith and holiness our souls are so contracted and dried up, that we cannot bear a sifting, but will go through the sieve, and perish with the chaff and base grain. We need holiness, brethren, for our own safety and happiness. And we need it that we may be successful 'workers together with God' in the great work of saving sinners. Look at the condition of the world lying in wickedness, and at least fifty souls go into eternity every minute. In the six minutes I have been speaking three hundred souls have gone to their long home. At least one hundred and fifty have doubtless gone to hell, and what are we doing to save poor dying sinners? O, it is a wonder that the angels' tears do not fall on us and wet our garments."

CHAPTER XI.

PREACHING IN A GAMBLING-HOUSE.

On the twenty-ninth day of January, 1851, a man called on me to attend the funeral of Charles B., a gambler, who, in a quarrel with a fellow-gambler the night preceding, was shot dead. "I think it a pity," said the man, "to bury the poor fellow without having some religious ceremonies said over him; and it will be a comfort to his friends."

He was laid out just where he was killed, in the "Parker House," on the east side of the Plaza. Taking my stand near the corpse, I sung:

"That awful day will surely come,
Th' appointed hour makes haste,
When I must stand before my Judge,
And pass the solemn test.

"Jesus, thou source of all my joys,
Thou ruler of my heart,
How could I bear to hear thy voice
Pronounce the word, "Depart!"

"The thunder of that awful word
Would so torment my ear,
'Twould tear my soul asunder, Lord,
With most tormenting fear.

"What, to be banish'd from my Lord,
And yet forbid to die?
To linger in eternal pain,
And death forever fly?

"O, wretched state of deep despair,
To see my God remove,
And fix my doleful station where
I must not taste his love!"

The singing and the occasion drew together nearly three hundred men, who stood uncovered before me. I announced as my text the last two verses of the book of Ecclesiastes: "Let us hear the conclusion of the whole matter: Fear God, and keep his commandments, for this is the whole duty of man. For God shall bring every work into judgment, with every secret thing, whether it be good, or whether it be evil." I then remarked as follows:

"Gentlemen, I always endeavor, in my public discourses, to adapt my remarks, so far as I can, to my audience. I take it for granted that the greater portion, if not all of you, are sportsmen; as such I shall address you.

"'The conclusion of the whole matter,' the great summary of life's duties, what is it? 'Fear God, and keep his commandments.' Do you understand it? You are not a set of ignoramuses. I know, from your appearance, that you have had early educational advantages. Some of you have had pious mothers to instruct you, and many of you, I doubt not, have been brought up in the Sabbath school, and you have all had the opportunity of reading the word of God, and of hearing it preached, from your boyhood to the present hour. You cannot plead ignorance. You know your duty: to 'keep his commandments.' How comprehensive the commandments of God, embracing every duty growing out of the relations we sustain to God and to each other! Had you given your hearts to God, believed in Jesus Christ, received the regenerating power of his grace in your souls, and were you, to-day, consecrated to his service, what happy men you would be! What an influence you might wield for God and his holy cause in California; help to build up good society, and to make this fair land a safe and happy home for your wives and children. The little boys and girls now growing up in our midst would repeat your names with grateful hearts, and call you blessed, when your bodies are beneath the ground, and your souls happy in the abode of angels and of God. But what are you about? What are you doing here in California? Look at that bloody corpse! What will his mother say? What will his sisters think of it? To die in a distant land, among strangers, is bad; to die unforgiven, suddenly, unexpectedly, is worse; to be shot down in a gambling-house, at the midnight hour—O, horrible! And yet this is the legitimate fruit of the excitement and dissipation, chagrin and disappointment, consequent upon your business; a business fatal to your best interests of body and soul, for time and for eternity.

"Again, look at its influence upon society. The unwary are decoyed and ruined. Little boys, charmed by your animating music, dazzled by the magnificent paraphernalia of your saloons, are enticed, corrupted, and destroyed, to the hopeless grief of their mothers, whose wailings will be entered against you in the book of God. Remember that 'for all these things God will bring you into judgment.' 'For God shall bring every work into judgment, with every secret thing, whether it be good or whether it be evil.'"

Every gambler listened with profound attention, and then formed the largest funeral procession, I believe, that I had, up to that time, ever witnessed in San Francisco. They returned, I presume, to their cards. One of them afterward said to a friend of mine: "That Plaza preacher is the strangest man I ever saw. He preached B.'s funeral, and said everything in this world he could think of against us, and yet he did not give us any chance to get hold of him!" He then paused a few moments, and, turning on his heel, said, "O J—s! didn't he give it to us?"

Five years afterward, when I was traveling in

the mountains, I was informed of two of the same gamblers, who had recently asserted that they never had been able to forget nor to shake off the impressions of truth made on their minds at B.'s funeral.

CHAPTER XII.

CITY OF SAN FRANCISCO IN AN UPROAR.

SUNDAY, twenty-third of February, 1851, was a day of great excitement in the city. It was ascertained that there was a large organized band of thieves and robbers in California in those days, operating at the same time in different parts of the state, yet all acting in concert. Men were knocked down and robbed in the streets, in the twilight; and stores and safes were broken open almost dayly. The night preceding the date above, a respectable clothing merchant, by the name of Janson, (now of the firm of Janson & Bond,) was knocked down behind his counter with a "slung shot;" and it was then thought that he could not recover. Two men, by the names of Windred and Stuart, were arrested on Sunday morning, and lodged in jail, as the supposed perpetrators of the deed.

The public forbearance, which had been taxed to the last point of endurance, now gave way to one almost universal burst of indignation. The people gathered round the jail to the number of about ten thousand men. I was requested by Windred's wife to visit him, as it was believed that the prisoners would be hung by the people before night. I had great difficulty in getting through the crowd; but finally succeeded in having an interview with the prisoners. Cries of, "Have them out! hang them!" etc., filled the air. It was with great difficulty that the public indignation could be suppressed, so as to give time for an examination and trial of any kind; but a doubt as to the guilt of the parties arrested, prevailed in allaying the excitement. I preached on the Plaza that day to about fifteen hundred persons, on the value and indispensable necessity of the Bible, believed in, practiced; indispensable to our safety and happiness, personally, collectively, socially, politically; the very foundation on which the glorious structure of our confederated nation is built; the chart by which we may navigate the stormy sea of life, and gain the peaceful haven of eternal rest. What does infidelity propose to do for us?

Good order and great seriousness prevailed. Windred afterward broke the jail, and cleared himself; Stuart was cleared by the courts. But the "Vigilance Committee of 1851" was organized as the result of these frequent robberies, and the inefficiency of the courts; and they executed some, and banished others, to parts unknown.

CHAPTER XIII.

CITY HOSPITAL ON FIRE.

How dreadful, in the stillness of the third watch of the night, is the cry, "Fire! fire! fire!" and the ringing of alarm-bells in all the wards of a large city. A livery stable full of horses in flames. Shocking! A mother and her infant in the third story of a building enveloped in fire; and the returning husband wringing his hands in phrensy. What a dreadful scene! Here, at the dead hour of night, a hospital, built of wood, on fire. It will consume to ashes in thirty minutes. In it are one hundred and thirty men—sick men—many of whom are unable to raise their heads from their pillows. No time for talk. Rush in, ye friends of suffering humanity. Let the strong men carry out the patients; take bed and all. Thus, in a few minutes, about half an acre of ground was strewed with mattresses, blankets, and dying men. The first thing was to get the sick off the damp ground on to the "cots," and provide covering to keep them from chilling to death in the night air. The next thing was to get some place of shelter. The "Waverly House," on Pacific-street, distant about half a mile, was offered. Many of the sufferers were immediately carried thither. But that required too much time. Next a two-story house was obtained, very conveniently located, but very inconveniently arranged. It had a narrow hall through the center, with narrow doors opening on each side into the rooms. A cot, containing a patient, could not be turned out of the hall through these doors; and hence we had to "unship" each patient in the hall, in order to twist him in through the doors into the rooms. All were rescued from the flames, with most of their bedding. Everything else pertaining to the hospital was consumed. The fire originated in a "house of the strange woman," adjoining the hospital. Most of our dreadful fires have started in some sink of iniquity. This fire occurred on the night of October 30, 1851.

A number of the patients were men who had been blown up, thirty-five hours before, in the explosion of the steamer "Sagamore." Some had broken limbs, and others were badly scalded. Some of them, on Monday, the twenty-eighth, had taken passage, in the City of Stockton, on board the steamer "Mariposi," to attend the celebration of the "admission of California into the Union," which took place on Tuesday, the twenty-ninth, in this city. On their way down, on Monday night, their boat was "run into" and sunk by the steamer "West Point," and they narrowly escaped a grave in the dark waters. On Tuesday afternoon, after the festivities of the celebration, they were blown up in the explosion of the "Sagamore," by which many lost their lives. On Wednesday night they were burned out, as above stated. My attention was called to an Irish boy, of about fourteen years, who lay on the ground in a dying condition. He said to his father, who stood by him: "Go away, go away from me. But for you I would not have come to this. You made me work, and drove me along for days, after I was so low with diarrhea that I was not able to work, and now I must die." The father tried to clear himself of the charges brought against him by his dying boy; but others present asserted that the boy told the truth. I tried to persuade the boy to forgive his father, and seek the forgiveness of his own sins through the merits of Jesus. A priest came along and took him in hand. What service he rendered him I know not. The boy died. As I carried in a poor Dane, who had been paralyzed, I saw Isaac Hillman (the razor-strop man) with a pot of warm coffee. I thought my Dane was dying, but a cup of warm coffee revived him. He has been in the City Hospital ever since, now about six years. No pen will ever delineate the sufferings I have witnessed in the hospital of this city.

CHAPTER XIV.

THE PREACHING THAT KILLED THE PLAZA CLOWN.

On Sunday afternoon, February 2, 1851, as I stood on the porch of the "Old Adobe," and sung up a thousand men, a good-looking fellow affected to act the clown. It was a clear, cool afternoon, but our clown came up with an old umbrella spread over him. In his right hand was a lantern, and in his left side-pocket a loaf of bread. Thus distinguished, after strutting round the circle of the audience, he came on the porch, near where I stood, lowered his umbrella, and tried to sing. I marked him in my mind, but said nothing. My text on the occasion was, "Let the wicked forsake his way." The first point was, Why should the wicked forsake his way? 1. Because the way of the wicked is exceedingly offensive to God. 2. It is most hideous and hateful in itself. Familiarity with it, and love for it, might blind and deceive us, but did not soften or change its nature.

> "Vice is a monster of so frightful mien,
> That to be hated needs but to be seen;
> Yet seen too oft, familiar with her face,
> We first endure, then pity, then embrace."

SINGING UP A CROWD AT THE OLD ADOBE ON THE PLAZA.

3. It is utterly ruinous in its effects to every interest of our souls, in time, in eternity. These points were duly illustrated and applied. One illustration used, showing how sin degraded the ennobling faculties with which God had endowed our souls, and disqualified us for the pure associations and spiritual delights of heaven, would be regarded by many persons as too ludicrous for a religious meeting; but the application was so direct on this occasion that the effect was good. It ran as follows: "On a trip to San José last week, in the steamer Star, our boat ran aground, and kept us there in the mud till after midnight. We had as passengers an alderman, a doctor, a general, a senator, a captain, and a high private, six high-minded, distinguished men, honorables of the land, noble spirits of the earth; none of your dull, sleepy fellows, you may be sure. (Colonel J. C. Fremont was aboard, but would not drink nor participate in any revelry.)

"While detained on the bar, they must have some appropriate enjoyment for the evening. The tastes and habits of such distinguished men furnish an example for all the boys of the land, and we should expect from such a source examples pure and elevating. Well, how did they spend the evening? The general said, 'Steward, have you got any good whisky?' 'Yes, sir.' 'Well, now, get us up a good bowl of whisky punch.' 'Ay, ay, sir.' The punch disposed of, they next played a game at cards. Then the alderman, who holds a chaplaincy in an association in this city, said: 'Steward, make us some more of that punch; it is first rate.' The table cleared again, they took another turn at the cards. Then the captain said, 'Steward, you are the finest-looking nigger I ever saw in my life; give us a little more punch.' After they had thus disposed of six bottles, they began to be very happy, and it was natural that their joyous emotions should find expression in song. God has endowed us with this talent of music, that by it we might express the joyous emotions of the heart, and sing his praise as the angels do. Now what do you suppose our worthies sung? They sung, over and over again, the song of 'Old Uncle Ned, with all the hair off his head.' Now, with angels and glorified souls, and all who have tastes adapting them to the enjoyments of heaven, the all-absorbing and soul-thrilling theme is the song of the world's risen Redeemer. But the highest point which the aspirations of these noble souls could reach, was the funeral dirge of a dead 'nigger.'"

The second division of the discourse exhibited the means of escape from the way of the wicked, urged by a variety of arguments. I took occasion to give the clown his "portion in due season," and when the Doxology was sung, he came to me trembling and and weeping, and said:

"Can you tell me what I am to do? I am a gambler and a drunkard, and a miserable sinner. I had a good mother, but she is dead, and I have no doubt that she is in heaven to-day. O, I am afraid there is no hope for me."

I took him by the hand and said: "If you go on in your present course, you will never see your mother again. But if you will quit gambling and drinking, and come out from your wicked associates, and attend church, read your Bible, and pray, and seek religion through the merits of Jesus Christ, you will yet be saved, and meet your mother in heaven. 'Let the wicked forsake his way.' Will you do it? Will you do it now? The Lord in mercy help you."

The poor fellow was greatly distressed, and I gave him a good deal of earnest talk about his soul, but I saw him no more. He probably, with half of my audience, left the city the next day for the mines. There are hundreds of men in the mines who have heard no preaching in California except what they have heard on the Plaza in this city.

CHAPTER XV.

HELPED TO A TEXT BY THE THIEF THAT STOLE MY MONEY.

On Saturday night, the fifteenth of February, 1851, I walked down town a few minutes, in company with Mrs. Taylor, on some business, and when we returned found that a thief had been into the house, and had opened trunks, turned over beds, and done a great amount of housework in the short time we had been absent. He stole, in money, about forty dollars. The next day, on the Plaza, I announced as my text the nineteenth and twentieth verses of the sixth chapter of Matthew. I read the text thus: "Lay up for yourselves treasures upon earth, where moth and rust doth corrupt, and where thieves break through and steal. But lay not up for yourselves treasures in heaven, where neither moth nor rust doth corrupt, and where thieves do not break through nor steal." I had a large audience, and they all looked as though they thought I had made a mistake; but I repeated the reading gravely: "Lay up for yourselves treasures upon earth," etc. Many looked at each other, and some whispered.

I then remarked, "Gentlemen, many of you were taught to read the Bible by your pious mothers, some of whom have since died in the faith, and gone home to heaven. You are all more or less acquainted with the teachings of this blessed book, although I am afraid you have not read it much since you came to California. The Lord have mercy on you. But you all conclude that the preacher made a mistake in reading the text. Now, I tell you, one of two things is true in regard to the matter: either, first, I have read it correctly, or, second, nine tenths of you are involved in a most shameful inconsistency of life; for you carry out, with the greatest possible earnestness, the teachings of the text as I have read it.

What is your business here in California? For what have you left your parents and friends, your wives and children, and braved the dangers of the deep, and of the desert? For what have you endured so much privation, and pain, and toil, in the rugged mountains of California? Is it all to 'lay up treasure in heaven, where moth and rust doth not corrupt, and where thieves do not break through nor steal?' Not a word of it. That has never entered into your purposes or plans. It constitutes no part of the object of your toils. All this privation, and peril, and suffering, and toil, is for the purpose of laying 'up treasures on earth, where moth and rust doth corrupt, and where thieves break through and steal.'"

The Spirit of the Lord was graciously present, and many sinners quaked under the discourse that followed.

CHAPTER XVI.

THE HUMAN HEART.

On Sunday, the twenty-seventh of April, 1851, at nine o'clock in the forenoon, I preached on the "Long Wharf," from the deck of the steamer "Union;" text: "The heart is deceitful above all things, and desperately wicked." Some of the arguments and illustrations used on the occasion were as follows: "The human heart may be compared to a jug, and why? Because we can only ascertain the character of its contents by what comes out of it. God is looking into your hearts now; but finite vision cannot penetrate the walls of that mysterious source of thought, and feeling, and action, which determines a man's character in the sight of God. But if we are allowed to judge of fountains by their streams, we have only to look at the foul streams of iniquity which continually flow through our streets to be assured of the character of their sources. See what profanity; what a desecration of God's holy day; what dreadful havoc is being made by that unrelenting slaughterer of human kind, the rum-seller; see what desolation is wrought in the city by the gambling fraternity; see the dreadful prostitution of female virtue; only behold the spirit of lasciviousness and covetousness, like the pall of death, spread over thirty thousand souls in this city! Our streets are thronged with God-hating, Christ-rejecting, pleasure-taking, sin-loving men and women. Remember, too, that these dreadful manifestations of the wickedness of the heart are but partial developments of its deep depravity, limited, First: By the restraints which are brought to bear on human conduct: social restraints, legal restraints, and religious restraints. Second: By the barriers of necessity, which circumscribe man's ability to execute the 'devices of his heart.' Look, for example, at that rum-seller. The house in which he lives, and from which are the issues of death, once belonged to a man of property and respectability. He lived there with his happy family; but the wily 'gentleman of the bar' took advantage of the moral imbecility of his victim, just as the highwayman takes advantage of the physical imbecility of the man he murders and robs. He has long since sent his victim's shattered, bloated carcass to a drunkard's grave, and his soul to a drunkard's hell. His family are in the 'poor-house,' dayly shedding fountains of tears more bitter than death. Now this is the business that man on the corner there follows. And why does he not treat every family so? Because he cannot.

His heart is guilty of the blood of every man he would decoy, if he could. So with the gambler; so with the swindling extortioner; and that swearer just out there in the street, who looks up and blasphemes the name of God, would, if he had the power, hurl him from his throne to-day. Yet such is the deceitfulness of the human heart, that, with all these diabolical volcanic fires pent up within, and the frequent outbursts of smoke, and flame, and flood, which spread moral desolation through the earth, and fill the world with woe; the possessors of such hearts will stand up in the temple of God, and, with all the apparent sanctimoniousness of an angel, say: 'God, I thank thee that I am not as other men.'"

CHAPTER XVII.

THE INDEPENDENCE BELL.

In the afternoon of Sunday, April 27, 1851, I had a large audience on the Plaza, to whom I said: "My text is recorded on the old Independence Bell, in the State House in Philadelphia, and reads as follows: 'Proclaim liberty throughout all the land, to all the inhabitants thereof. Lev. xxv, 10. By order of the Assembly of the Province of Pennsylvania, for the State House in Philadelphia. Pass & Stow. Philadelphia, A. D. 1753.' Young America, just beginning to scribble, thus wrote his name on that old bell, twenty-three years before the tocsin of war called him forth to try his manly muscles in mortal combat with a giant foe.

"On the evening of the great atonement of the Jews, the Jubilee year, that proclamation, sounding from every hill-top in Palestine, and echoing through every vale from Dan to Beersheba, thrilled with gladness the hearts of millions of Abraham's sons and daughters. It was under the inspiration of the Bible doctrine contained in this text, that John Hancock and his compatriots were enabled, with steady hand, and a determination of purpose stronger than death, to sign that immortal document, the 'Declaration of Independence.' This was the theme that clothed our fathers with that unconquerable courage and zeal, which carried them through a seven years' struggle on fields of carnage and blood, till throughout the united colonies, from Maine to South Carolina, the jubilee trump sounded.

"Behold, to-day, the results of an appropriate, practical application of Bible truth, even politically and civilly considered! But the institution of the Jubilee typifies a spiritual Jubilee, which, in its provisions and results, transcends all earthly good and earthly glory, as much as the duration and developments of eternity transcend the duration and developments of time. Our divine Joshua proclaims: 'The Spirit of the Lord is upon me, because he hath anointed me to preach the Gospel to the poor. He hath sent me to heal the broken-hearted, to preach deliverance to the captives, and recovering of sight to the blind: to set at liberty them that are bruised. To preach the acceptable year of the Lord.' And he hath sent forth his heralds, charging them to 'Go into all the world, and preach the Gospel to every creature.' This is the character in which I appear before you to-day.

"'Sent by my Lord, on you I call;
The invitation is to all:
Come all the world! come, sinner, thou!
All things in Christ are ready now.

"'Come, all ye souls by sin oppress'd,
Ye restless wand'rers after rest;
Ye poor, and maim'd, and halt, and blind,
In Christ a hearty welcome find.'"

CHAPTER XVIII.

KING DAVID'S FOOL.

My Plaza text for Sunday, March 2, 1851, was: "The fool hath said in his heart, There is no God." Some of my remarks on that occasion ran as follows:

"Here is a watch my father gave me when I was a boy," holding it in my hand. "He bought it from an old bachelor by the name of Walkup, who, of course, recommended it to be a first-rate watch. I am not acquainted with its early history, but if I were to tell you that this watch had no maker, that some happy chance formed the different parts of its ingenious machinery, and that another chance put them together with the very useful design of a time-piece, you would call me a fool. It is said that Sir Isaac Newton had a friend who professed to be an atheist. Sir Isaac, anticipating a visit from his friend, placed a beautiful new globe where he knew it would arrest the attention of his visitor. When the atheist saw it, he exclaimed with admiration, 'Sir Isaac, who made this beautiful globe?' 'O, it was not made at all, sir!' answered the great philosopher, with a significant glance at the confused eye of his friend. The argument was unanswerable. And if we cannot believe that a mere globe of wood, with certain lines, and colors, and figures, representing the earth's surface, could come by chance, how can we imagine that this mighty globe itself, with its continents and seas, and various laws, to say nothing of the vast universe of suns and systems which occupy the immeasurable expanse of space, could be the result of chance? To adopt such a conclusion, there is surely no such fool in this intelligent audience. But remember, David's fool was not such. He probably, like some fools encountered by Jesus, in the days of his incarnation, 'drew nigh to God with his lips, and honored him with his mouth,' but 'said in his heart, There is no God.' The Holy Spirit was looking at him, and heard his heart say it, and moved the royal Psalmist to pen it down, and so it stands recorded to-day.

"The Holy Spirit is looking at each one of you now, and listening to every pulsation of your moral heart, and were he now to reveal what has there passed this day, what shocking revelations he would make! It is not by the profession of the mouth, but by the conduct of men, that we are to learn the orthodoxy of their hearts. A miserable gambler said to me but a short time since, 'When I came to California I had but twenty-five cents; but I had good luck playing cards, and by and by set up a "monte-table," and, I thank God, I have been very successful.' He said he was a member of the Roman Catholic Church, and professed to be very devout.

"A wretched rum-seller over here on Jackson-street, had filched the pockets of a poor fellow, wrecked his constitution, blighted all his hopes for time and eternity, unstrung his nervous system, and driven him into delirium tremens; and when his poor victim was dying, the tender-hearted rum-seller, full of sympathy for the suffering, sent in haste for me to come and pray for the poor man.

"Why, these gamblers round the Plaza here, whenever they shoot a fellow, go right off for a preacher to pray over their dead. One who came for me to preach at the funeral of C. B., who had been shot the night before just there in that large saloon, said, 'We thought it would be a pity to bury the man without some religious ceremonies. It will be a comfort to his friends, too, to know that he had a decent Christian burial.'

"I have buried three such within as many months. They profess a belief in God, but their conduct gives the lie to their profession.

"What is the swearer's notion of God? Even to-day my ears have been saluted with the horrid oath. They do not believe in their hearts that there is a God, and but use His name in ironical contempt, or else they have so degraded a notion of God as to treat him worse than they would treat a dog. They would not think of so treating a fellow-man. 'The fool hath said in his heart, There is no God,' but every pulsation of that heart gives the lie to the blasphemous assertion. This system of bones, and sinews, and muscles, and arteries, and veins, and nerves, so fearfully and wonderfully made, proclaims, 'There is a God.' And this still more mysterious soul, which occupies this highly-wrought tenement, proclaims yet more loudly, 'There is a God.'

"See him in nature. See him in his providential government over men. See him as revealed in his word. See his mercy—his justice. We belong to him. To him shall we answer for all the sayings of our hearts. Do you believe in him? Do you obey him? Do you love him? Are you on friendly terms with him to-day? If not, 'We pray you in Christ's stead, be ye reconciled to God.' Will you sue for pardon and reconciliation now?" etc.

After the benediction, a stranger spoke out, saying, "Gentlemen, you all know how laboriously and successfully Father Taylor labors here on the Plaza from Sabbath to Sabbath. Now I move that we take up a collection. I will not urge you to give; I know you are all ready."

"Pass along the hat," said one.

"Let it come this way," said another.

'Stop, stop," said I. "Gentlemen, I am much obliged for your kind feelings, but I never allow a collection to be taken up out doors for my benefit. I have preaching every Sabbath twice in the church on Powell-street, and all who are so disposed can give there; but you will please do nothing of the kind here. I cannot have my street preaching trammeled by collections."

I have now preached (July, 1856) about six hundred times in these streets; have occasionally taken up collections for poor men and for building the Bethel, (I collected $400 at one time on the Plaza for the Bethel,) but have never taken up one collection for my own benefit, though often in need. My reason is, that in the streets I proclaim a free Gospel, "the royal proclamation," to heathens and Christians, to Jews and Gentiles, to Catholics and Protestants, to inhabitants of every nation, and I am unwilling to furnish ground for any of these to impugn my motives, or to say, "He can afford to sing and preach in the streets when he gets a good collection every time."

The Lord, in pity, remember thy unworthy servant in the "day of thy coming."

CHAPTER XIX.

THE TIME THE LORD DID NOT "KEEP THE CITY," AND WHY—THE GREAT FIRE.

At eleven o'clock in the night of Saturday, May 3d, 1851, a fire broke out in our city, which raged till nine o'clock in the forenoon of Sunday, the fourth. It was the most destructive fire by which this city has ever been visited. The loss was variously estimated from twelve to twenty millions of dollars. Several hundred passengers had just arrived on the steamship New Orleans, on the evening the fire occurred, and the city was filled with strangers besides, so that it was impossible to tell how many persons perished in the conflagration. The ashes, it was believed, of six men were found in the ruins of T.'s iron building. It was said that five of them rushed in to rescue a sick man, who was confined to his bed inside, and when they got back to the door, it was so warped by the heat that they could not open it, and the fire in the street was so great that it was impossible to relieve them. And there they perished, at the threshold of life.

GREAT MAY FIRE OF 1851.

Many of the streets were planked, and on each side were wooden sewers, which served as flues to conduct the fire, and greatly facilitated its destructive progress through the city. Our "Old Adobe" escaped, and at the appointed hour for preaching, I stood in my place on the "porch." It appeared to be a very unpropitious time for collecting an audience. The people were running to and fro, under a high pressure of confused excitement, and many were busy in collecting together their little savings from the fire, many tons of which were scattered in tangled confusion all over the Plaza. I, however, threw out, amid the smoke, and dust, and noise, of the vast field of desolation which was spread out before me, one of Zion's sweetest songs, and drew together about one thousand men. My text on the occasion was, "Except the Lord build the house, they labor in vain that build it. Except the Lord keep the city, the watchman waketh but in vain." "Are we to understand, my hearers, from this text, that it is unnecessary to employ builders or watchmen? Certainly not. But having them, and using all the appliances necessary to build up and preserve our city, we must, nevertheless, rely, for success and safety, upon the merciful Providence of God. 'Except the Lord build the house, they labor in *vain* that build it. Except the *Lord keep the city*, the watchman *waketh* but in *vain.*'

"When we consider the numerous causes and occasions of fire, the millions of smoking cigar stumps scattered over the city, and how many thousands are recklessly careless in the use of fire, and how many hundreds of malicious spirits, who are, from motives of revenge or a desire to pillage, always ready to fire the city, the wonder is, that the city is not fired every week. And why is it not? Because the Lord keeps the city. He overrules these occasions, either by preventing their application, or by arresting their progress in time to save the city. It is not necessary for him to work miracles to effect this. Having absolute control of all the forces and agencies in the universe, save the internal moral exercises of the human will, he can bring his purposes to pass by any of the agencies of human care and precaution, or by the so-called accidents which parenthetically intersperse the whole drama of life. As I walked out the other day, I very opportunely saw the kindling of a great fire, by the careless throwing of embers among shavings; but for the accidental discovery, that fire might have swept over the city. 'Well,' says one, 'if that is the doctrine, and the Lord is the keeper of the city, how is it that our city is now burned up, scarcely anything left but the smoking ruins of her greatness of yesterday?' Let us inquire whether there may not be reasons why the Lord should, at certain times, make an exception to this general rule of his preserving providence. As we are rational and moral agents, he deals with us on moral principles. Those reasons, therefore, must be sought by an examination of our conduct, as subjects of his moral government. The Lord has been very kind to us in the past; kind to us individually, and kind to our city collectively. This you cannot deny. But how have we requited his kindness? Just look abroad through the city, as it was yesterday. See what a wholesale desecration of God's holy day. As many as seven hundred places of business are open in this city every Sabbath day. Look at the rum traffic and its deadly effects; think of the fornication and adultery practised in the city; hundreds of men, too, frequenting those haunts of infamy, who have confiding wives and interesting children at home. Do you imagine that God is an indifferent spectator of these diabolical scenes? Listen to the horrid oaths which continually ring through our streets. I said to a swearer this morning: 'Be patient, my friend, and don't swear about it.' 'Patience! patience! talk about patience,' said he, 'and the city burning up!' 'Well, sir,' said I, 'but what good does it do to swear about it?' 'Ah,' said he, 'it does to let the gas off.' Now what kind of gas is that which smokes and bubbles in the hearts of so many thousands of men in this city, the 'letting off' of which consists in the foulest blasphemy against God? It is this awful gas, my friends, which has kindled and fed the flames which have consumed the city. It is the gas of carnal enmity against God, manifesting itself in so many horrid forms in our midst, that 'breaks the bands of God asunder, and casts his cords from us.' And when we break the moral 'bands' that bind us to God, we, by the same violence, break the providential 'bands' that bind God, in the plenitude of his mercy and providential care, to us. Let the citizens of San Francisco beware! God is dealing with them. This disaster, dreadful as it appears to be, is but a premonition of 'judgment to come,' in consequence of their sins. It is also a disciplinary measure for the correction and improvement of our morals. Now, if you wish to become loyal subjects of our Divine Sovereign, and rebuild the city on a permanent and safe basis, you must have that dreadful gas removed from your hearts. The only remedy by which it can be neutralized and extinguished is the blood of the crucified Jesus. The fire could not be put out last night because of the scarcity of water, and the inefficiency of the means to apply it; but 'the fountain opened in the house of David for sin and uncleanness,' is abundant, free, and available.

"'Its streams the whole creation reach,
So plenteous is the store;
Enough for all, enough for each,
Enough for evermore.'

"'The blood of Jesus Christ, his Son, cleanseth us from all sin,' and the Holy Ghost is here to-day to apply it to your hearts. Will you accept of the remedy? Will you?"

The application of the discourse and the exhortation that followed were, we believe, attended by the unction of the Holy Spirit. The occasion throughout was one of great solemnity. Many hearts throbbed with emotion, and many eyes were filled with weeping. "Except the Lord" apply the word, they labor in vain that preach it.

CHAPTER XX.

ARRIVAL OF MISSIONARIES.

ON Sunday, the fifth of November, 1851, I preached at nine o'clock in the morning on the Long Wharf, from the character and conduct of the patriarch Jacob. At half past ten I preached on Pacific Wharf, from "Jacob wrestling with the angel of the covenant." While preaching on this occasion, the steamship "California" arrived. Among her passengers were T. H. Pearne and lady, missionaries, *en route* to Oregon. Many passengers arriving in those days had the impression that they would hear no preaching in California; then to hear, the first thing, even before they disembarked, the appeals of Gospel truth was, to many, a matter of inexpressible surprise.

That afternoon I preached on the Plaza, from Hosea iv, 2: "By swearing, and lying, and killing, and stealing, and committing adultery, they break out, and blood toucheth blood. Therefore shall the land mourn, and every one that dwelleth therein shall languish." I was enabled, through the unction of the Holy Spirit, to deal very plainly with all the characters implicated by the text. I learned from Brother Pearne, who was a hearer at our Plaza meeting, that A. Bland, my brother-in-law, a missionary to California, his wife and child, were on the steamship Republic, then due. I soon afterward heard that the Republic was wrecked on a sunken reef about twenty miles from the "Golden Gate," and that the rush of water through the leak had put out the fires, and that she was in great danger of going down, with a large freight of human beings. I waited the news of the fate of my friends with no ordinary degree of solicitude. The steamship California was immediately dispatched for the relief of the wrecked vessel, which, though leaking very badly, was kept up by the pumps and buckets used by the passengers, till she was towed in on Monday afternoon. The passengers, though frightened and weary, were all safely landed. How many souls have been wrecked just outside the harbor of eternal blessedness!

"Jesus, lover of my soul,
  Let me to thy bosom fly,
While the nearer waters roll,
  While the tempest still is high;
Hide me, O my Saviour, hide,
  Till the storm of life is past;
Safe into the haven guide,
  O receive my soul at last."

CHAPTER XXI.

THE WHISKY-BARREL PULPIT.

In September, 1851, one Sabbath morning, on Pacific-street Wharf, I asked Captain L. for permission to preach from the deck of his steamer, but he respectfully declined granting the favor, saying, "There are some men at work aboard, and I am afraid it would interrupt them." I then took a position close by, so that I could give the captain and his men "a portion in due season," and to the crowd as well. I happened to get for my pulpit on that occasion a barrel of whisky, (I have preached probably a hundred times on the heads of liquor barrels,) which stood on the wharf, and prefaced my discourse by saying, "Gentlemen, I have for my pulpit to-day, as you see, a barrel of whisky. I presume this is the first time this barrel has ever been appropriated to a useful purpose. The 'critter' contained in it will do me no harm while I keep it under my feet. And let me say now to you all, to sailors and to landsmen, never let the 'critter' get above your feet. Keep it *under your feet*, and you have nothing to fear from it."

At the close of the sermon the congregation gave me a collection of one hundred and twenty dollars toward the erection of our "Bethel."

THE PORK-BARREL PULPIT.

The Sabbath following I occupied as a pulpit, at the same place, a barrel of pork. I remarked, as I balanced myself on the head of the barrel, "I see my pulpit of last Sabbath, the barrel of whisky, is gone, and I am very much afraid that my timely warning, as is too often the case, was not heeded, and that its contents have ere this gone down the throats of some of our fellow-citizens. I have in its stead to-day, as you see, a barrel of pork, literally less of the spirit and more of the flesh. But this is God's house while I here dispense his word, as really as the spot where Jacob slept and dreamed, and saw the ladder that reached up to heaven. God was in that place, and God is here this morning. Jacob's God is looking at you now. O that the Spirit of his grace may this hour subdue your fleshly lusts, while I deliver to you a message from him who sent me."

My text on this occasion was from Proverbs, third chapter, thirteenth and fourteenth verses, "Happy is the man that findeth wisdom, and the man that getteth understanding. For the merchandise of it is better than the merchandise of silver, and the gain thereof than fine gold." On which I made the following remarks:

"This inestimable treasure wisdom, what is it? 'She is a tree of life to them that lay hold upon her,' verse eighteenth. The very 'tree of life,' from which our first parents were driven, and from which they were debarred because of sin, by the 'cherubim and a flaming sword which turned every way to keep the way of the tree of life.' It is the favor of God. It is reconciliation through the blood of Jesus. It is experimental religion. We do not possess this treasure naturally, nor do we acquire it intuitively, nor without *earnest* effort. Jesus has propitiated the throne of Divine justice, and opened the gateway to the tree of life, and says to a world of outsiders, 'Seek to enter in at the strait gate.' 'Strive to enter in at the strait gate; for many, I say unto you, shall seek to enter in, and shall not be able.' If we would find 'wisdom,' we must 'seek for her as for hid treasures.' Many of you are just down from the mines. You have 'made your pile,' and now you are on your way with hearts beating with hopeful emotion, to see the friends you love. But if you should find a watery grave on your voyage, how you will need religion. Above all things else, be sure to seek and lay in a good supply of it before you embark. But we were going to ask you how you got your gold. Did you not have to seek for it, and dig deep and toil hard to get it? You were impelled, in your diligent search, by desire, and hope, and faith, and determination, and patience. So must you seek if you would obtain religion. True, our 'works' do not constitute a meritorious ground of our acceptance with God, but an indispensable condition, on which God, for Christ's sake, graciously imparts salvation to our sin-stricken hearts. The miner says, 'Happy is the man that findeth gold, and gets ready to go home to his friends.' We say, upon the authority of God, 'Happy is the man that findeth wisdom,' and gets ready to go to his home in heaven, to meet his friends who have gone before him. Again, 'Happy is the man that getteth,' or, as it reads in the margin, 'that draweth out understanding.' As 'wisdom' is the attainment of the best ends, by the use of the best means, so 'understanding' is the *fruit* and *experience* of wisdom. When a soul is 'regenerated,' it receives the principle of spiritual life, as in natural generation it received the *principle* of *natural* life. Now the development of this principle of spiritual life in the heart, and its corresponding manifestation in the life, is what is meant by 'drawing out understanding.' 'Happy is the man' that retains and develops his religion. Now some of you, after having made your 'pile,' have been decoyed into the gambler's hell, and have, in one short hour, lost the labor of years. So many of you, who were once so happy as to find wisdom, having failed to draw out understanding, have been decoyed by the god of this world, and robbed of your treasure. California is full of backsliders, and they are the most miserable men, and many of them the meanest men in this land. They are of two classes: first, those who, before they left home, adopted that fallacy of the devil, that 'It is impossible to live religion in California, and therefore it is no use to try.' One old apostate said: 'God don't hold any man to answer for his conduct, after he crosses the Missouri River.' And thousands have staked the interests of their souls on that lie. Another, who, it is said, was a preacher once, said: 'I knew I could not carry my religion through California, so when I left my home in Missouri, I hung my religious cloak on my gate post, till I should return.' Thus, if he ever had any religion, he threw it away before he started for California. This is the worst class of backsliders. They backslide in principle, deliberately.

"The second class embraces those who came to California fully determined to live to God; but they wandered away into the mountains, where they were cut off from all the privileges of the sanctuary, and association with Christians. There they became feeble, got discouraged, and were finally 'entangled and overcome.' Poor fellows, prisoners of war! The

Lord have mercy upon them! Jesus is looking after you, my backslidden brother, as he looked after apostate Peter. He is very anxious to save you; and he will save you, if you let him. Will you? The backsliders of both classes are unhappy: the whole of them. But, blessed be God, we have the men here in California, who, in opposition to flame, and flood, and death, have 'drawn out understanding,' and they are 'happy.' 'The merchandise,' or exchange value and circulation of this article, namely, developed religion, 'understanding drawn out,' is 'better than the merchandise of silver, and the gain thereof than fine gold;' even 'fine Yuba gold.' By the 'gain of gold, and the merchandise of silver,' you may make sunshine friends; supply the wants of your mortal bodies, which will be dead and rotten in a few years; gratify your fleshly lusts, which will, when the sources of all gratification are cut off, as they will be when your tabernacle is taken down, like so many vultures, prey upon your deathless spirit forever. Your money, to be sure, may be applied to useful purposes. It will buy you a cabin ticket to New-York; but it will not secure you even a *steerage* passage across death's dark flood. It will give you position among the honorables of the land; but it will not secure you the favor of God and good angels. It will build a church, if you please; but it will not buy your soul a place in heaven. A man who came to California in 1848, and made a fortune, laid him down, not long since, in Washington-street, in this city, and died. He had plenty of silver and gold; but, as he informed me, was destitute of religion. When dying, he said: 'It is very hard. I have just got ready to live; and now I must die.' What a miserably poor man he was. An old colored man, from Baltimore City, died recently, in the City Hospital, on Pacific-street, but a few blocks from this spot. He was a very homely old man, and suffered intensely with the 'king's evil,' and I don't know how many other evils, and had not one red cent with which to bless himself; but he had 'wisdom,' and was 'happy.' I saw him frequently, and every time he was happy. A short time before his death I administered to him the sacrament of the Lord's Supper, after which he clapped his bony black hands, and shouted the praise of God. Said he: 'The Lord only knows how I have been pinched with poverty, and what this poor body has suffered; but I am rich. I have an inheritance in heaven, glory be to God! I shall soon be released from these sufferings, and go to my home in heaven;' and then the good old 'darkey' sang, just as the colored people alone can sing:

"'A home in heaven! as the sufferer lies
On his bed of pain, and uplifts his eyes

To that bright home, what a joy is given,
With the blessed thought of his home in heaven.

"'A home in heaven! when our pleasures fade,
And our wealth and fame in the dust are laid;
And strength decays, and our health is riven,
We are happy still with our home in heaven.

"'A home in heaven! when the faint heart bleeds,
By the Spirit's stroke, for its evil deeds;
O! then, what bliss, in that heart forgiven,
Does the hope inspire of a home in heaven.

"'A home in heaven! when our friends have fled
To the cheerless gloom of the moldering dead;
We wait in hope on the promise given,
We will meet up there in our home in heaven.'

"I wish you could have seen how his big eyes glistened with rapturous delight, as he thus sung of his 'home in heaven.' Religion gave him a royal heirship in the kingdom of glory.

"The truth of our text he proved in life, confirmed it by his triumphs in death, and is now realizing it in the fruition of a blessed immortality in heaven.

"Now, my friends, you see the prize; you have heard the price; if you like the terms, close to-day. Will you do it? Will you do it *now?* 'Behold! now is the accepted time. Behold! now is the day of salvation.'"

A French grape dealer, and a Spanish pear and orange seller, had each a movable stand on which they exposed their fruits for sale wherever they thought they could get the greatest number of customers. Seeing me draw together great crowds of men each Sabbath, they thought it would be a fine thing for them to patronize street preaching; and while I was administering to my audience "the bread of life," they would improve the repast by adding a little good fruit. So one bright Sunday morning in November, 1851, when I went to my appointment on Pacific Wharf, there they were, in front of my "barrel pulpit," the two stands side by side, with their fruit arranged in the most inviting style. I mounted the "barrel" without appearing to notice them, and sung up a congregation. While I was singing the first verse of "The Old Family Bible," a man cried out, "Where did you get your Bible?" Just at that moment I was ready to commence the second verse, and sung to him,

"'The Bible, the volume of *God's inspiration.*'

"That's where I got it, sir; by the inspiration of God," and then sung on.

"'At morning and evening could yield us delight,
And the prayer of our sire was a sweet invocation,
For mercy by day, and for safety through night.
Our hymns of thanksgiving with harmony swelling,
All warm from the hearts of the family band,
Half raised us from earth to that rapturous dwelling
Described in the Bible that lay on the stand.

"'The old-fashion'd Bible, the dear, blessed Bible,
The family Bible, which lay on the stand.'

"Had you been blessed, my friend, with such a sire, and had you been trained in such a family band, you would not ask me where I got the Bible;" and sung on:

"'Ye scenes of tranquillity, long have we parted,
My hope's almost gone, and my parents no more;
In sorrow and sadness, I live broken-hearted,
And wander unknown on *this far-distant shore.*
Yet how can I doubt a dear Saviour's protection,
Forgetful of gifts from his bountiful hand?
O let me with patience receive his correction,
And think of the Bible that lay on the stand.

"'The old-fashion'd Bible,'" etc.

The audience now stood in a circle about twenty deep, as close as possible, and the fruit dealers in the center. I then said, "Grapes, pears, and oranges! Gentlemen, you must not suppose that I have any interest in this Sunday traffic in calling you together around it. I hope you will not patronize these Sabbath-breakers. You are not so grape-hungry but that you can wait till to-morrow, and then during the six days in the week lay in a supply for Sunday. These fellows have set up here, expecting to make a fine speculation out of my audience this morning; but they will find that they have brought their fruit to the wrong market."

The fruit dealers by this time would gladly have got out of the range of our artillery, but they were completely environed; and I gave them grapes gratuitously, and pared them down to the smallest point my pity would allow me. I then preached from Jacob's dying address to his sons, and a blessed season we had.

The poor Spaniard and his French neighbor, like the Shechemites of old, did not understand the refined arts of modern times for making religion subservient to mercenary purposes. They, however, did not miss it so much as did a grocer I heard of, who, for a long time, paid a high pew-rent in a certain church in this city, and afterward complained, in his simplicity, saying, "Now, for so many months I have paid my pew-rent in that church, and I and my wife always went in just as the congregation turned to face the choir, so that I know they could not help seeing us, and I don't believe that it has benefited my business one cent. Not one of them comes to buy at my store."

The Lord pity such miserable sinners, who make a "stalking horse" of religion. The possessor of gains thus acquired, will have more trouble with them than Rachel had with her stolen gods, and in the end share a worse fate than Hamor and Shechem.

CHAPTER XXII.

WAYSIDE HEARERS.

ONE Sunday morning in October, 1851, I preached to a large audience on "Long Wharf," from the parable of the Sower. Illustrating how "Satan cometh immediately, and taketh away the word that was sown in their hearts," I said of his Satanic majesty, that "Just at the moment the good seed would take effect, he excites in the heart of the hearer opposing passions, or diverts his attention by presenting to his mind some attractive scheme or train of thought, while he devours the seed; or by sending a wagon-load of calves through the midst of the audience, to the great annoyance of attentive listeners." (A load of calves for the market at that moment was passing through the crowd.)

The audience so blocked the street sometimes from side to side with a living mass of humanity that it was difficult for a man to get through. A wagon or dray would therefore be subjected to considerable delay in making a passage through, and I frequently took advantage of the opportunity, and gave them a little "grape" as they passed. Once when a lean-looking man, driving a poor horse, was trying to urge his way through the crowd, I said, "Look at that poor man! Working seven days in the week is bringing him rapidly down to his grave! A man cannot break the law of the Sabbath without violating a law of his own constitution. Look at his sunken, sallow cheeks, and his dim eyes! How the sin of Sabbath-breaking is telling on him! He'll die soon if he don't reform. Look at his poor old horse! The Lord ordained a Sabbath for that horse, but his merciless master is cheating him out of it. See there, how he beats him. After all, I had rather be the horse than the man, if he dies as he lives."

On another occasion a wag, thinking to have a little sport, tried to ride through the crowd on one of the smallest of that small species of animals, the Jack. His animal refusing to go through, I said, "See there, that animal, like Balaam's of the same kind, has more respect for the worship of God than his master, who only lacks the ears of being the greater ass of the two."

The man, in great confusion, beat his animal out of sight in double quick time. The reader may wonder how I managed to restore the equilibrium of the audience after such a scene. I always tried to anticipate that difficulty, and would follow such scenes by the most solemn appeal the subject in hand would allow. The sudden surprise of such appeals sometimes produces a thrilling effect for good. An important end is accomplished when a sleepy congregation is by any legitimate means fairly waked up. First melt, and then mold the metal.

CHAPTER XXIII.

A MOTHER'S TODDY-LOVING SON.

A MOTHER, to whom God intrusted an infant heir of immortality, a beautiful boy, with instructions to train him for holiness and heaven, dosed her dear little boy with sweetened toddy, and taught him early to be a wine-bibber. No doubt she feels great solicitude for the welfare of her son since he left for California, and as she has not heard from him for a long time, for he seldom writes, I will give her a word of information concerning him. He has not been to church in San Francisco, for he was not taught to go to church even at home, and is not likely to form such a habit here. But he passed by where I was preaching one bright Sunday morning, in the spring of 1852, on Pacific-street. He listened a while, as most passers-by do, but he had been indulging a little, and was not in a good condition to receive the truth. After meeting, I saw him before me as I walked down Sansome-street. He "fetched up" in front of a large liquor-store, where a cask of brandy lay with a little pump in the bung. He looked for a moment with great apparent interest at the cask, as though he thought it a rare opportunity for cheap rations, and his gestures seemed to say, "O, for a demijohn!"

But, mother, you know your son is a smart, inventive youth, as you used often to tell him, when his wits were sharpened by your sweetened toddy. So he immediately hit on the following happy expedient. Taking off his hat, he pumped it full of brandy; and as, with joyful steps, he bore away his prize, every now and then he stopped and dipped his red nose into his hat. When I came "along side," I leaned over to smell the contents of his hat, so as not to be mistaken in my facts, and your generous son said to me: "Come, take a drink, won't you?" Not fancying the article, nor the vessel containing it, I respectfully declined. Your dear boy was well provided for that day, and probably got a good night's lodging on a free ticket, in the station-house. I have not seen the precious youth since, unless, by possibility, he were the same man that I saw soon after in the bay. He had been "fished up" by some boatman, and was tied by one of his legs to a "pile," to await the arrival of the coroner, whose jury gravely sits on such cases, and at the city's expense, returns a verdict of "accidental drowning."

CHAPTER XXIV.

THE DEATH OF BELSHAZZAR.

On Sunday morning, January 4, 1852, I stood on the deck of the steamer Webber, at Long Wharf, and announced as my text: "In that very night was Belshazzar, king of the Chaldeans, slain." Nearly opposite to where I stood, on the other side of the wharf, lay the steamer Empire, which had been chartered to convey a company of California legislators on that day to Vallejo, the seat of the legislature of this state at that time. The Empire was steaming up for her Sunday excursion, while I was trying to raise the steam on the Webber against Sunday excursions. My song drew to the side of our boat a large crowd, while the embarkation of the honorable legislators drew an equally large crowd to their boat, but I had the whole of both parties within the compass of my voice, and I preached to the Empire party more especially. As I doubted whether many of them ever went to church, I thought it a rare opportunity for giving them a little Gospel truth.

I illustrated, by the life of Belshazzar, that a Sab-

LAGER-BEER POLITICIANS.

bath-breaking, licentious, carousing, drunken man, was utterly unfit for any official position in the gift of any respectable nation; and to elect men to make our laws, whose brains were addled with brandy, and who showed so little respect for one of the highest laws and most venerable institutions of God, the holy Sabbath, was a wicked absurdity and a burning shame to the American people. I did not design, by these reflections, to implicate the whole of the California legislature, for it contained some very good men, but I thought them peculiarly applicable to the party addressed on that occasion. I illustrated further, the end of such a course of procedure, by the Mene, Tekel, Peres, the numbering, weighing, and dividing of the Chaldean kingdom, and the slaying of her wicked king. Already we begin to see the handwriting of doom on the wall of our illustrious palace of American liberty. God has given us a glorious country,

"The land of the free and the home of the brave."

But let the American people beware! God is the author of all our blessings, and must be honored.

A number of months after this occasion, a stranger called on me, and requested a private interview. Said he to me: "Do you remember preaching from the deck of a steamboat at Long Wharf, nine months ago, from a text concerning the destruction of Babylon, and the death of Belshazzar?"

"I preach there every Sunday morning. O yes," I replied, "I do remember it now, by the Sunday excursion which started that morning from the opposite side of the wharf."

"That was the time to which I allude," said he; and then related the following facts concerning himself: "I was up to that morning a confirmed Universalist; and was withal a very wicked sinner. As I was walking leisurely down the wharf that morning, I heard you singing, and went into the crowd, through curiosity, to hear what was to be said on the occasion. While you were preaching, a strange fearfulness, which I cannot describe, came over me. I felt a smothering sensation at my heart, and thought I was dying. My Universalism all vanished like smoke; and I felt that if I died then, I should certainly go to hell. For some time I knew not what to do. I came very near crying out; but something seemed to say to me, 'Pray, pray to God, in the name of Jesus Christ, for pardon.' So I began earnestly to pray. For three weeks I suffered a constant fearfulness and trembling. I felt every moment as though some dreadful calamity or judgment was about to befall me. I was afraid to go to sleep at night, lest I should wake up in hell; and every day there seemed to be literally a heavy mist before my eyes, which made everything look dark and dreary. But all these three dreadful weeks I continued to pray; and suddenly, while I was praying, and trying to trust in Jesus Christ, it appeared to me that a stream of light shone right down from heaven into my heart, and in a moment I realized that my burden of sin was gone; and instead of fearfulness, and a nervous tremor, I felt all the vigor of renewed youth. The mist of my eyes gave way to the brightness of morning. I praised God for his pardoning mercy. I have been up in the mountains ever since. I have had but few public religious privileges, but have had my private prayers, and have been recommending religion to all my associates. Jesus has been very precious to my soul all the time. To-morrow I expect to embark for China; and I wanted to see you before leaving, and have a talk, and get some tracts and religious books for distribution aboard ship. I feel as though I ought to do all I can in the cause of Christ, for his great mercy to me, and for the great desire he has given me to see poor wandering souls converted."

He did not expect soon, if ever, to return to California. So we closed our interview with a final farewell, and a mutual pledge to each other to live for God, and meet again on the other side of the river.

CHAPTER XXV.

A PERSONAL COLLISION ON THE PLAZA.

SUNDAY afternoon, January 18, 1852, I preached on the Plaza, from the judgment recorded on one side of Zechariah's "flying roll:" "And every one that sweareth shall be cut off, as on that side, according to it." Zechariah v, 3. Subject of discourse, profane swearing, a dreadful sin, notoriously prevalent in California. Near my feet sat a stout, muscular man, who, though not drunk exactly, had rum enough in him to make him impudent. When I read the text, he looked up and said,

"Now go ahead, sir."

I proceeded accordingly; and in a few minutes he said, with an oath,

"You are a fine-looking fellow; I want to have your lithograph taken, sir."

Said I, in an undertone, "My friend, you must be quiet, or you will have to go away."

"Who will take me away?" said he boisterously. "I would like to see the man, or any two men, who could take me away. Let any man touch me, if he dare."

Contrary to all precedent in my experience, before or since, no one tried to remove him. He seemed to have intimidated the people, who, as I afterward learned, were waiting for a policeman for whom they had sent. I, however, had reached a point in my discourse, where I wanted to illustrate the insinuating progress, and even increasingly degrading effects of the habit of profane swearing, and said: "Here, gentlemen, you see my subject illustrated. This is a man of fair talents and good education; a man who might be a very useful member of society in California, did he but follow the advice of his pious old mother. He once had a tender conscience, and a good reputation. The first oath he ever swore alarmed him, and he promised that he never would swear another; but he came off here to California, and fell into bad company, and now look at him. Step by step he has gone down the road of degradation, and here he is to-day—the holy Sabbath day—half drunk; and here to disturb, by his foul oaths and curses, a worshiping assembly. What can be done for such a case? My text says, 'He shall be cut off.' But it is a dreadful thing to see him sink under the judgment of God, and bring down the gray hairs of his old father in sorrow to the grave. The Lord have mercy on him."

I had never seen the man before; but learned afterward that I had hit it exactly in every particular of his history. He sat perfectly quiet while I took his "daguerreotype;" but after a few minutes he rallied, and said,

"I want you to get away from here, and let me talk." In a moment he sprang to his feet, and as I did not give back, there was a collision. He took hold of my coat collar, and I seized his arm, and gave him such a shaking as muscles, developed at a currier's beam, can give, and passed him to a couple of men, saying, "Here, lead this fellow away. Don't hurt him, but take him out of sight."

"Now," said I to the audience, "while they are disposing of that fellow I'll sing you a song of Zion;" and then I sung two or three verses of

"Forever here my rest shall be,
Close to thy bleeding side;
This all my hope and all my plea,
For me the Saviour died."

The encounter greatly increased the audience, and I took up the subject where I left off, and proceeded with the discourse. The meeting from that to the close was unusually interesting. I give a single point as a specimen of the manner in which a crowd of swearers were addressed on the occasion: "But the swearer will say, 'I have got so much in the habit of swearing that I cannot quit it. I have often tried, but it's no use.' Indeed! Is that so? If you have come to that, just own up now, as honest men, that you are the bond slaves of sin and children of Satan, 'led captive by the devil at his will;' bound down by the chains of habit to a practice which must exclude you from heaven, disqualify you for decent society, and eventually shut you up in hell. O pitiable condition! An angel would weep to see your souls, endowed with attributes capable of almost an unlimited development of intellectual strength and moral excellence, thus bound in chains to the galley oars of the devil. Upon your own confession, my dear sirs, this is your moral condition to-day. O let your cry ascend, 'Who shall deliver me from this dreadful bondage of death?'"

In regard to the spirit in which I encountered my antagonist, I will here insert an extract from my journal, written immediately after the occurrence took place:

"To one who did not see the circumstances, my action in taking hold of the fellow may seem rash. I have this to say, that it was done in the midst of a sermon, and in almost as short a time as it would take a man to clear his throat. I think I felt no other spirit than that of preaching my sermon out, in spite of the devil or any of his agents. In resisting the man I had no feeling of ill-will or design of injury, but simply to effect what I succeeded in doing. I confess there was nothing premeditated about it; but I feel thankful to the Lord that I was preserved from bad temper, and that no one present, as I believe, suspected me of any other design than the one expressed. Though the man got a good griping and I got my coat torn, the result was only greatly to increase the multitude of attentive hearers, with no one to disturb them. Mr. Hall, the United States Commissioner of Public Buildings in California, who was present, told me that L. was, at home, a very respectable and sober man, and one of the best mechanics in the country; but, with too many others, had fallen in California. Poor fellow! I am sorry for him. O Lord, pluck him as a brand from the fire."

CHAPTER XXVI.

A LIVING ILLUSTRATION.

In the spring of 1852, as I was on the Long Wharf one Sunday morning, discoursing to a large audience on the "one thing needful," I proceeded first to show that it was needful to the well-being of the bodies of men. That religion, as a regulator of the appetites and passions, preserved men from a great variety of excesses which were destructive to health and happiness. Illustrating this, I said to the crowd: "Go with me, if you please, through the hospitals of our city. Ask the hundreds of sufferers to whom I will introduce you, the cause of their afflictions; and, while you will see some good men, brought down by unavoidable diseases, you will find that a large majority of those miserable beings have been there imprisoned for the violation of physical laws, from which this needful thing would have saved them."

"That's true, Dr. Taylor; that's true, every word of it," cried an old man in the audience.

"Yes, sir," said I, in reply; "you know it by sad experience. There, friends," I continued, "you have a living illustration of the truth of my position.

"That old man, lacking this needful thing, indulged his appetite for strong drink, and, as a consequence, I found him two years ago in the hospital. He lay there for many months, suffering everything, but death. The physician succeeded at last in doctoring up his old carcass, and if he had given his heart to the Lord, and obtained the healthful, preserving influence of his grace, he might have continued a well man. But he went out still destitute of the one thing needful, and in a short time he again took the cup of death, for which he had to serve another long term in the hospital. With naturally a good constitution, if he had been possessed of vital godliness, the probability is, he would not have lost a day from sickness in California. He is a ship-master, and capable of doing well for himself and his family; and he came here, too, at a time when he had a good cpportunity to make a fortune, and but for the want of this one needful thing, he might to-day be reclining on his well-earned California fortune, by his own happy fireside, surrounded by the wife of his youth and the lovely children the Lord has given them. But look at him. Here he is, a mere wreck of manly strength, found ering on the lee-shore of the dreadful sea of inebriation; his wife clad in habiliments of mourning, blacker than the widow's weeds, and his beautiful daughters disgraced, poverty stricken, and broken hearted. I fear he will never see them again, and if he does, he is unfit for the relations, duties, and associations of the head of such a family. (The poor old captain was now weeping and crying audibly, as a boy that was being castigated.) I would not, my friends, unnecessarily hurt the feelings of the poor old man. He knows I am one of the best friends he has in this land, and that I have often entreated him as a brother, and prayed at his side, and have done everything in my power to keep him from self-destruction, and to induce him to seek the one thing needful."

In the next place I went on to show, by a variety of proofs and illustrations, the value of religion to the soul.

CHAPTER XXVII.

CALIFORNIA HUSBANDS MEETING THEIR WIVES.

THE darkest chapter in the history of California is that which records the disruption of family ties and connubial relationships, occasioned, primarily, by the rage and rush of thousands of heads of families to her mines of gold. Many families of children have been thus neglected when they most needed a father's watchful care and counsels. Many a wife has pined with a broken heart on account of the absence of her husband, and the husband a desolate, isolated wanderer in a strange land. In very many cases, these husbands are unsuccessful, and often unable even to raise money enough to carry them to their poor, dependent families at home. Very many of both husbands and wives have died without the longed-for "meeting again." The mails, surcharged with death shocks, have for years been passing back and forth, from ocean to ocean, and ever and anon, suddenly and unexpectedly as a thunderbolt from a clear sky, the lightning leaps from the train and strikes the widow's heart, and hope is gone. Still more dark and dreadful is the record of connubial infidelity, which has hopelessly sundered and desolated hundreds of once happy families.

In the midst of all these dangers the meeting of true and faithful husbands and wives, after weary years of separation, is an occasion of thrilling interest, and often furnishes scenes which baffle the painter's skill. Such scenes occur at our wharves on the arrival of each ocean steamer. A few incidents characterizing them are contained in the following extract from my journal.

"*Tuesday, Feb.* 3, 1852.—I boarded the steamer Panama upon her arrival this afternoon, to see if there were any missionaries aboard. Her trip had extended three days beyond her time, and much solicitude was felt for the safety of her precious freight of five hundred passengers.

"About four thousand persons crowded down Long Wharf to witness her arrival. Quite a company of anxious wives, who had come to join their husbands, stood on deck, looking out to catch in the distance the joyful recognitions of those they loved. One simple-hearted, beautiful little woman, getting a glimpse of her husband in the crowd, clapped her hands, and danced for very gladness. One man rushed on deck, and threw both arms round his wife, as though he would run right away with her, and then, with arms around each other, they walked 'abaft' in the greatest glee, not seeming to be conscious that anybody was in sight of them. Nearly all that met embraced and kissed each other, some laughing and some weeping, amid the cheering of the multitude. A Mrs. Gardner, who had less of youthful fire than many, but I should say not less of genuine affection, was quietly seated on deck, waiting the arrival of her husband. The old gentleman took off his hat when he got within a few feet of her, and with his venerable bald head bared, approached her with an air of dignified affection which I cannot describe."

On another occasion a man of my acquaintance, by the name of Brown, who was expecting the arrival of his wife, pressed through the crowd with eager haste to see if she was aboard, and inquired:

"Is Mrs. Brown aboard?"

"Yes," answered one, "she is down in her room."

She, in the meantime, learned that Mr. Brown was coming, and was filled with raptures at the thought. Mr. Brown found the room, and rushed in to embrace his dear wife, and, to their mutual disappointment, it wasn't either of them. He was not the man, and she was not the woman.

But a sad case I saw, and it was one of many of the same kind. A man hasted aboard with joyous heart to meet his wife, and was told that three days out from Panama, she had suddenly sickened and died, and had found a grave in the deep blue sea. He was taken to her state-room, and there were her trunks and clothing, and everything just as her own hands had left them. Ah! the sadness of that heart has never been told.

When all the ship Zion's company shall have arrived in her destined haven, what joyful meetings we shall see on Canaan's shore. Shall our friends, who have gone before us, find us aboard, or shall it be told them that at such a time we left the ship and were drowned in perdition? The Lord bless my readers, and help them to prepare for the happy greeting on the other shore.

CHAPTER XXVIII.

"DECLARATION OF INDEPENDENCE."

On the fourth of July, 1852, I preached a temperance sermon on the Plaza. I drew a parallel between the oppressions of our fathers and mothers, under the administration of King George and his train of high officials, and the more dreadful sufferings of tens of thousands of our fellow-citizens, under the despotism of King Alcohol and his long train of officers, thousands of whom are quartered in our midst and pampered at our expense. I drew a picture of the aggressive marches of the enemy, and the horrible havoc he was making of American flesh and blood, and property, and tenderest ties, and dearest hopes, and asked them what they would do if any foreign potentate or power should invade our territory and commit such outrages with the bayonet. Shades of Patrick Henry! Wouldn't Uncle Sam's boys rally and run to the rescue? "Come forward to-day, like John Hancock and his invincible compatriots, and sign this 'Declaration of Independence.'" About forty persons came forward and signed the temperance pledge. While I was discoursing, an old woman, who kept a grog-shop, close by where I stood, came out and cried:

"Don't listen to him. He's an impostor. He's preaching for money. He's telling you lies."

"Dry up, old woman," replied some of the outsiders; "dry up! We know what's the matter with you. Your craft is in danger. He is taking away your customers. We know Father Taylor. He's a good man, and he's telling the truth, and nothing but the truth." The woman immediately disappeared. Just as I closed my remarks, a man tried to get the attention of the audience, and said: "This man is an impostor, hallooing round here to get people's money." "Stop, stranger," said one; "what is your business here in the city?" "Why, sir," replied the fellow, after being closely pressed for an answer, "I am a gambler, and I did a first-rate business, and made money here, till these preachers came to the city. But this fellow is hallooing at the people here every Sunday, and has broke up my business. I can't get a decent living." "Good! good!" said one and another. "Hearken, friends," said I; "this gambler has paid me a high compliment. He says I have broken up his business." "Good! good!" responded the people, the gambler suddenly "vamosed," and I have not laid eyes on him since.

CHAPTER XXIX.

PROFANE SWEARING.

SUNDAY afternoon, July 27, 1852, I had, on the Plaza, a congregation of about eight hundred. The subject of discourse was profane swearing. The application of the sermon ran thus:

"Swearer, what do you mean? Do you thus use the name of God ironically, to express your utmost contempt of the idea of a God? Then you are an atheist. What a vast herd of atheists we have in our midst, who dayly feast on the acorns of Divine beneficence, and never acknowledge the oak whence they fell. Or do you believe that God takes pleasure in unrighteousness, and looks approvingly upon your corruption and wickedness? Then you are a Mohammedan. What a host of Mohammedans we have in this Christian country. 'No, we are not Mohammedans, nor are we atheists; we believe in a wise, good, and holy God, but we believe he is too good to damn a soul for a few years of sin.' Then you do not believe the Bible, and you are a set of infidels. What a crowd of infidels we have in this land of Bibles.

'No,' says one, 'I believe the Bible, I believe in God and in all his attributes, in Jesus Christ and the Holy Spirit, and I believe in future rewards and punishments, in heaven and in hell.' That, then, swearer, is your faith, is it? You are a pretty orthodox fellow after all. But then, with this faith you stand up here, in the light of God's holy day, and with eyes, and face, and voice, and gestures, all indicating the greatest possible earnestness, you, in the presence of witnesses, most solemnly call upon God to 'damn your soul to hell.' How little did the Persian of whom the poet speaks, know of the import of such prayers:

"'A Persian, humble servant of the sun,
Who, though devout, of bigotry had none,
Hearing a lawyer, grave in his address,
With adjuration every word impress,
Thought the man a bishop, or, at least,
God's name so oft upon his lips, a priest;
Bow'd at the close, with all his graceful airs,
And ask'd an interest in his frequent prayers.'

"Had the prayer been explained to the poor Persian, he would have fled from it as from a boa constrictor. What does such a prayer imply? There is a man who has a family at home. He has long been trying to get ready to return to them, and his wife and children are constantly writing him, 'Husband, do come home.' 'Father, we want to see you so badly we can hardly live; when will you come home?'

You would like to see them all, would you not? And yet, as you walk these streets, you are dayly praying, and in your prayers, asking God with great apparent earnestness, that you may be cut off, and never allowed again to see your wife and children, and that your wife be left a desolate widow, and your children helpless orphans, to mourn over you, as those who have no hope, and that your business here may be left in other hands, and that your poor carcass may be hauled out and stowed away unwept, with the rotting thousands of Yerba Buena Cemetery, and that your soul, covered with guilt and shame, may be 'damned eternally by God.' O horrible! Such a prayer chills my blood. Yet these are the petitions of the largest proportion of my praying audience. Well, do not be discouraged; your prayers will be answered in due time. God is a prayer-hearing and a prayer-answering God, fulfilling all his promises, and his promise to you is, 'Every one that sweareth shall be cut off.'"

CHAPTER XXX.

A SABBATH DAY'S WORK.

An extract from my journal:

"*November* 1, 1852.—Yesterday morning, (Sunday,) I, as usual, led class at 9 A. M. Preached on the Long Wharf at 10 A. M. Preached in the Bethel at 11 A. M. Preached in the State Marine Hospital at half past two P. M., and on the Plaza at 4 P. M. Preached again in the Bethel at 7 P. M. These are my regular appointments for each Sabbath. It may seem to be too much for one man to do, but the Lord gives me strength to do it, without inconvenience or injury to myself. O Lord, give me moral power and good success in soul saving.

"I will here record my thanksgiving to the Lord for the rescue of my little boy, Morgan Stuart, who fell into the bay this morning, where the water was ten feet deep. O Lord, I thank thee, that when my dear boy was sinking into the deep, thou didst enable me, at a single leap, to seize him and bear him above the surface of the foaming tide, till timely and safely conveyed ashore. Glory be to my merciful God."

At a class-meeting held in the Bethel, in November, 1852, a seeker of religion arose and said, "On my passage to California a fellow-passenger died just before we reached Acapulco; and after the captain had given orders to heave him overboard, some one found about his body a paper, certifying that he was an Odd Fellow. The Odd Fellows aboard at once countermanded the order of the captain, and after proper arrangements, took him ashore and buried him respectably, with the honors of an Odd Fellow. A passenger said to me, 'Now I see the necessity of being an Odd Fellow.' Said I to him in reply: 'Now I see the necessity of being a Christian, and by the grace of God I will seek religion.' I have been seeking religion ever since, and am determined never to give up the struggle."

CHAPTER XXXI.

"SAVE ME FROM MY FRIENDS."

On Sunday afternoon, October 2, 1853, I was preaching on the Plaza to about a thousand hearers from the text, "And he, trembling and astonished, said, Lord, what wilt thou have me to do?" In the midst of the discourse a drunken fellow began to mutter, and tried to create a disturbance, when another, pretty well intoxicated also, said to me:

"Captain, I hope you will not consider it an interruption, and I will beat this contemptible fellow like ——. I won't allow a preacher, a good man like Father Taylor, to be interrupted in the discharge of his duties. No, that I won't," clinching his fist at the same time, and making a push toward his antagonist.

"Stop, my friend," said I. "I know you are my friend, and that you want to preserve order; but that fellow is pretty badly scared now, and will, I have no doubt, remain perfectly quiet without a whipping. If he will not, then it will be time enough 'to pitch into him.' Just hold on now, if you please."

"Very well," said he; "I'll do just as you tell me. If he don't behave, I am on hand to give him the heaviest licking he ever got in all his born days."

My friend, a man I never saw before nor since, had already made a great deal more disturbance than the foe, but quiet was very soon restored, and the preached word was manifestly attended by the Spirit of the Lord. At the close I sung Bishop Hedding's hymn:

"Ye angels who mortals attend,
  And minister comfort in woe,
Come listen, ye heavenly friends,
  My happier story to know.
I sing of a theme most sublime,
  No sorrow my song can control:
I sing of the rapturous time,
  When Jesus spoke peace to my soul," etc.

I took the wounded from the Plaza to the Bethel and that night we had eleven persons at the altar for prayers, three of whom then, and the rest soon after, professed to experience pardoning grace.

Meeting a young man in the street, he thus addressed me: "How are you, captain? I know you; I heard you preach on the Plaza. I encouraged you then; I contributed toward building the Bethel. I was sober then, and respected God. Now I am drunk, now I respect the devil." The devil elicits a great amount of respect by the wholesale and retail rum traffic, in which he is so extensively engaged.

CHAPTER XXXII.

DEFENSE OF THE SABBATH.

In January, 1853, an article appeared in the "Alta California," a popular dayly of this city, over the signature of "Merchant," against the Sabbath as a day of religious observance. He attempted to prove, from the Hebrew Bible, that nothing more was contemplated in the institution of the Sabbath than a day of recreation, feasting, and dancing. He announced that that was the first of a series of articles on the same subject. The Sabbath following, January 30, I had a large audience on Long Wharf, and took my text from "Merchant's" article in the newspaper, and preached on the origin and design of the Sabbath. The merchant, unhappily for himself, had chosen Nehemiah as his favorite author, so we sent Nehemiah after him to deal with him, as he did with "the merchants and sellers of all kind of ware" which he expelled from the city of Jerusalem, for doing "as these Long Wharf merchants do here every Sunday." How successful I was in presenting the truth, and in "showing up" the fallacy of "Merchant's" positions, could, perhaps, better be decided by the congregation in attendance. But the rest of "Merchant's" series on the same subject never appeared. By the way, I had the pleasure of numbering our good Bishop Ames among my auditors on that occasion. Our street congregations usually stand up, but I honored our good bishop with a seat on a pile of wood which lay on the side of the wharf; and I will be pardoned for the liberty I take in saying, that he looked as good-natured and maintained his dignity as creditably to himself on that pile of wood as I have ever seen a bishop in his chair in Conference.

CHAPTER XXXIII.

EVIL TIDINGS.

The following extract from my journal was written, as per date, from the best information I could get from passengers who were witnesses to the scene described, and to newspaper reports:

"*Friday, April* 1, 1853.—Yesterday was a day of evil tidings. We had an arrival of the surviving passengers of the wrecked steamship Independence.

"The wreck occurred February 16, on Margarita Island. After striking a reef, which caused a very bad leak, a sail was drawn over the broken part, and the ship was headed in for the shore, for a convenient place to beach her. They succeeded, after a run of four miles, in grounding her, but the breakers were very heavy, and the first boat sent ashore with a line, swamped. The second boat succeeded in carrying a hawser ashore; but by this time the water had so risen in the ship as to stop the lower flues, and throw the fire out of the furnace doors, and in a moment the ship was in flames. An indescribably horrible scene ensued. Hundreds jumped overboard, and many, on floating spars and other light material, which had been thrown over for the purpose, succeeded in reaching the shore. A poor sick man asked to be carried to the bulwarks of the vessel; then crawled overboard, and sunk to rise no more. Rich men offered large sums to be taken ashore, but there was none to help. Mothers were seen running after their children, and one by one throwing them overboard. One poor mother chased her last frightened child to the verge of the flames, as it fled from her, and caught it and threw it into the ocean, and then jumped over to sink with her children beneath the dark waters. A Brother Knox, of precious memory in our church at Sacramento City, there found a watery grave. A Captain Taylor conceived the idea that he would take his child between his teeth and his wife under his arm, and breast the breakers, and make the shore. He accordingly lowered his wife by a rope at the stern of the ship, and told her to hold on till he should come with the child; but in the meantime, some one threw the child over, and he lost it. He then took his wife and swam ashore. She, however, from the loss of her child, and unavoidable injuries and exposures, took fever, and died just before they reached this port. She was buried in this city. The hawser, it is said, was hanging full of persons, who were working their way toward the shore, when one of the officers, fearing it would break and drop its load, ordered it slackened a little; but, unhappily, it dipped and submerged the whole line of strugglers after life beneath the breakers. Out of between four and five hundred passengers, it was believed that about one hundred and fifty perished.

"In addition to the above sad intelligence, I learned on yesterday that Brother J. Benham, an educated and promising young man from Brooklyn, New York, who was admitted on trial into the California Conference, at its recent session, and sent to Sacramento River Circuit, was drowned a few days since in attempting to ford Catche Creek, when very high. He was a young man of bright promise, and was, I learn, succeeding on his circuit very well indeed. How strange that a career so promising should close so early and so suddenly. How very important that we be always ready."

We are horrified at the details of those dreadful disasters, which by one fell swoop carry, as in a moment, multitudes of our fellows from our midst into eternity. But we are reminded that the bodies of all our race are under sentence of death; that awful sentence incurred by sin, "Dust thou art, and unto dust shalt thou return." And every thirty years sweeps off a generation, each one suffering as much, in the aggregate, probably, as those who perished in the wreck of the Independence. When they drop off one by one, but very few, besides the immediate friends of the deceased, in any given case, feel the shock. And even the kindrod themselves gradually prepare their minds for the event, so that equal horrors with those of a steamship disaster, are so distributed as to lose, in a great degree, their startling effect. While no one dies because of a foreordained decree, neither does any one die by chance; nor until, in view of his moral relations to God as a probationer, he is summoned from the stage of life by the Great Supreme. When a cause for such a summons exists in our moral relations to God, we, having fulfilled the mission of life and ripened for glory, or having filled up the measure of iniquity, and made ourselves "vessels of wrath fitted for destruction," then, any one of the ten thousand occasions of death by which we are surrounded, is sufficient to push us off the stage. The difference between the different modes of these occasions or secondary causes is very slight. Now, in the order of Providence, a great many of such persons, and none others, are hurried simultaneously into eternity, for the warning and moral improvement of the living. "For when thy judgments are in the earth, the inhabitants of the world will learn righteousness." "But," says one, "if that doctrine be true, why should I be at any trouble to preserve or restore health? I shall die anyhow when my time comes." But, sir, the violation of the laws of health, and the neglect of all available restoratives, are sins, which affect those moral relations of which we speak, just as any other sins do, and by your sins you may class yourself with those of whom the Psalmist says, "Bloody and deceitful men shall not live out *half their days*."

CHAPTER XXXIV.

THE REPROBATE SAILOR REDEEMED.

On Sunday afternoon, June 26, 1853, I found a man in my Bible-class, who seemed to be in distress, and I engaged him in the following conversation. Said he, in answer to my inquiries:

"I was educated in my youth for a Universalist preacher, but I could not believe the doctrine, and instead of preaching I went to sea. I believe in the doctrine of foreordination and reprobation. I have been in great distress of mind for fourteen years. My soul is all over diseased. I have had no peace except what I got by drinking. I drank rum to relieve my distress. I have been hoping that God would have pity on me, and bring me in; but I fear he never will do it. I fear I am a reprobate, and that there is no hope for me."

"But, my brother," replied I, "God has declared, in the most solemn and unequivocal manner, 'As I live, saith the Lord God, I have no pleasure in the death of the wicked, but that the wicked turn from his way and live. Turn ye, turn ye from your evil ways: for why will ye die?' Again, it is a declaration of inspired truth that Jesus Christ, 'by the grace of God, hath tasted death for every man.' What for? Did he make a mock provision for such as were reprobated to eternal death?"

"Ah, but we are told," said he, "that though 'many are called, but few are chosen.'"

"Truly; but does God call the 'many,' and proclaim to them the tidings of salvation deceitfully, to mock their fears and aggravate their bondage under chains of inexorable fate? Surely the righteous God is sincere in his offers of mercy to all sinners. Christ answers the question, why so 'few are chosen' of the 'many called.' 'Ye will not come unto me, that ye might have life.' Now, my brother, God has been very desirous to save you for a long time; but you would not let him. He has been calling you for fourteen years, and you would not come. Instead of hearkening to the voice Divine and obeying your Lord, you ran off to a grog shop and got drunk. Do you ever pray to God for mercy?"

"What!" said he; "I pray! I pray! Why it would be blasphemy for such a wretch as I am to pray. 'The prayers of the wicked are abomination to the Lord.'"

I replied, "Solomon says, 'The *sacrifice* of the wicked is abomination;' but it is nowhere said in the Bible that the prayers of a penitent sinner are abomination; but it is said, 'Let the wicked forsake his way, and the unrighteous man his thoughts: and let him return unto the Lord, and he will have mercy upon him, and to our God, for he will abundantly pardon.' How is he to do this? By 'calling upon God while he is near.' He is near you now, my brother. The agony of soul you feel, and these tears, prove that his Spirit is now operating on your heart The Psalmist, as a poor sinner, cried to God from the horrible pit. Says he: 'I waited patiently for the Lord; and he inclined unto me, and heard my cry. He brought me up also [in answer to that cry] out of a horrible pit, out of the miry clay, and set my feet upon a rock, and established my goings. And he hath put a new song in my mouth, even praise unto our God.' And so the poor publican, who felt as guilty as you do, and 'could not lift up so much as his eyes unto heaven, but smote upon his breast, saying, God be merciful to me a sinner! I tell you this man went down to his home justified,' pardoned in answer to a sinner's prayer."

"O, but," said he, "they were not near so bad as I am. The iniquities of my fathers for four generations seem to be visited upon me."

"O, you know," said I, "that the proverb, 'The fathers have eaten sour grapes, and the children's teeth are set on edge,' has passed away long ago, so far as answering for the *sins* of our fathers is concerned. Within the last fortnight more than half a dozen sinners, equally as bad as you, some of them the worst men in the city, have, in this Bethel, called upon God and obtained mercy, and they are happy in his love to-day."

So soon as the Sunday school and Bible class closed, he was taken into the shipkeeper's room, where, surrounded by some warm-hearted sailors, he cried to God, in the name of Jesus, and in an hour experienced "redemption through the blood of the Lamb, even the forgiveness of his sins." That afternoon, after preaching on the Long Wharf, he went round with a bundle of tracts for distribution, and manifested great zeal in trying to persuade his fellow-seamen to "ship" for the celestial port. He soon afterward went to sea. The Lord keep him steadfast.

CHAPTER XXXV.

THE DRUNKEN SUICIDE'S FUNERAL.

On the twenty-sixth of May, 1853, I attended the funeral of W., of Pennsylvania, who had the previous night committed suicide by the use of laudanum. He lay in a small, filthy shanty, attended by ten of his bar-room companions. The undertaker had not arrived when I entered the shanty, but the friends, in their generous haste, proceeded at once to put down the lid of the coffin.

"Good-by, Bill," said one, as he fitted the coffin-lid, and then they went to work to set the screws. One used an old razor; another an old knife; two others employed themselves in pressing in the coffin and fitting the screws; a fifth went off in haste to borrow a screw-driver, that the work, as he said, "might be finished up decently."

In the meantime I proposed to them the following question: "How did this man come to his death?"

"Hard drink," said one. "I've known him here for three years. Hard drink was the thing, sir."

"No," said another, "Bill was one of the best

boys in this city. He had his failing, and would drink, as we all do, but he was a first-rate fellow."

"It was a sore face," said a third, "which pained him so that he got disheartened and took laudanum."

"No," said the fourth, "it was a punishment. He could not help it." (He meant it was so decreed.)

"Well," said yet another, "I think it was his misfortune. He was driving a dray in the city, and had bad luck, and got discouraged, and put an end to himself."

I then arose and sung:

"That awful day will surely come,
The appointed hour makes haste,
When I must stand before my Judge,
And pass the solemn test," etc.

I then said: "It is a solemn thing to die. To die in our sins is dreadful; but for a man to rush, by the violence of his own hands, unbidden, into the presence of a sin-avenging God, is too horrible to be described. What could lead this man to such a dreadful end?" I then quoted their testimony on the subject, and continued, "If this man had been a praying, sober man, would he have had that 'sore face?' If he had 'been diligent in business, fervent in spirit, serving the Lord,' would he probably have had such 'hard luck?' and, if so, would these two evils combined have led him to destroy himself? Now the facts in the case are these: The 'sore face,' the 'hard luck,' the discouragement and depression of spirits, were all the results of his drunkenness. And 'hard drink,' as this man has truly said, was the sole cause of his death. Now, how did he become so 'hard a drinker?' By tippling. When he used to drink, as you say you all do, he did not dream of such an end. Thus the fatal habit grew on him. Do you not know that the chains of habit are stronger than chains of steel? You are every day forging chains which bind you down more and more tightly to an infamous destiny.

"Why do you drink? Because it gratifies your vitiated appetite. Every repetition, as you imagine, increases the gratification. The absence of this gratification creates, as a man said to me one day after preaching on the Long Wharf, 'a terrible pain down in here,' which must be relieved. And thus, by the combined forces of the pleasure and the pain occasioned by the absence of it, your desire for the deadly cup becomes more and more imperative.

"This is the philosophy of this ruinous habit. 'A failing.' Ah! it is a fatal 'failing!' You cannot imagine where it will lead you. Your only hope of a better end than the case of this poor man, is to 'taste not, touch not, handle not the unclean thing.' Begin now to pray, and cry to God in the name of Jesus for mercy, to forgive the past, and for grace to cure this ruinous habit, and to preserve you in the future. The Lord have mercy on your poor souls. I hope you will take warning from the awful end of your friend, and the dreadful grief it will occasion his widowed mother. For the sake of your bodies, and for the sake of your property, and for the sake of your blood-bought souls, and for the sake of your friends at home, tarry not, fly as from the mouth of hell, and lay hold of the hope set before you in the Gospel. It is your only hope for time and for eternity."

The fellows gave most serious attention, and really seemed to feel the force of truth, and a "desire to flee the wrath to come."

CHAPTER XXXVI.

A PEEP INTO A CALIFORNIA LOVE-FEAST.

THE following is a description of a love-feast, held in the Seamen's Bethel, San Francisco, May, 1853: After a hearty, united song of praise to God, and prayer by the Rev. I. Owen, presiding elder, and the tokens of mutual Christian affection, between thirty and forty witnesses of Jesus testified to his saving love. Brother R. said: "At Buenos Ayres, in South America, God found me a poor sinner. I was, as too many seamen are, a most profane swearer. The Spirit of God mightily convinced me of sin. I thought of my mother's prayers and tears, and how I had lived. I wept and cried to God for mercy; and there he spoke peace to my soul. Thank God for a praying mother. When I returned to the State of Maine, to tell my friends what God had done for me, I learned that on the very night God converted my soul at Buenos Ayres, my mother and other Christian friends were at a prayer-meeting, and prayed specially for me. Glory be to God, the prayers of the righteous avail much! I enjoyed the life and power of religion for several years. But when I arrived in California, in 1850, from a long voyage at sea, I was in a backslidden state; but I inquired for a Methodist Church, and went up the first Sunday to the chapel on Powell-street. After preaching, Brother Taylor invited seekers of religion and backsliders to come to the altar. I went right forward; and there God again spoke peace to my soul. For three years he has kept me in California by his grace. I have 'shipped for the run,' and intend to 'keep my course,' till I land my soul in Canaan's peaceful haven."

Then all sung:

"Never more will I stray,
From my Saviour away,
  But will follow my Lord till I die.
I will take up my cross,
And count all things but dross,
  Till I meet my Redeemer on high."

Brother B—t said: "I love the Lord Jesus Christ, for he sought me at a camp-meeting away up in Iowa, and spoke peace to my soul. In the midst of great trials, and much unfaithfulness on my part, he has led me along for ten years. I feel his love burning in my heart to-day. Glory be to his holy name!"

The following song seemed to speak in melodious strains the feelings of every heart:

"O, Jesus, my Saviour, in thee I am blest!
My life and my treasure, my joy and my rest!
Thy grace be my theme, and thy name be my song,
Thy love doth inspire my heart and my tongue.

"O, who is like Jesus! he's Salem's bright King;
He smiles and he loves me, he taught me to sing:
I'll praise him, I'll praise him, and bow to his will,
While rivers of pleasure my spirit do fill!"

Brother L—g said: "I left Germany when a poy, and came to New-York. I heard Brother Lyon preach in New-York, from de text, 'Ye must pe porn again.' He came strange to me. I tid not know vot to do. I vent again, and kept going, till one Sunday morning the Lord converted my soul. I am still on my vay to Canaan's happy shore."

Then the song was sung:

"O, Canaan, bright Canaan! I'm bound for the land of Canaan," etc.

Brother L—d arose, exclaiming: "Canaan is just where I want to go. At the corner of Pacific and Battery streets, in this city, God converted my soul. I came to California a very wicked sinner; but the first Sunday after I landed I heard Brother Taylor preach on the wharf and on the Plaza, I was struck under deep conviction; and for three weeks afterward a more miserable man than I was never walked these streets. I have heard a great deal about taking up claims, and about Spanish titles, since I came to California. Brethren and sisters, I have staked off my claim, which I intend to hold and work forever. There's no Spanish title covering it. The devil, the most incorrigible old squatter that ever visited this world, has tried to 'jump my claim,' and dispossess me; but I resisted him in the name of the Lord, and he fled away. I have a clear title to 'an inheritance incorruptible, undefiled, and that fadeth not away, reserved in heaven for me.'"

As he sat down, the audience sung:

> "There is a land of pure delight,
> Where saints immortal reign;
> Infinite day excludes the night,
> And pleasures banish pain."

Brother S—r said: "I feel that I am indeed in a love-feast. Here we are from almost every part of the world. We have taken a little bread and water as tokens of our mutual love; and we feel the love of Christ flowing into our hearts, uniting us all in one bundle of love. Brethren, my soul is happy."

Brother C—r said: "At our last quarterly meeting I determined to live nearer to God. I feel that I have been better. I have many precious seasons all alone with Jesus, aboard my vessel. Four years ago. in old Massachusetts, on the eleventh day of March, God spoke peace to my soul. I went to church one night swearing, determined to break up the meeting. The devil said he would help me; but when I got to church, the preacher met me at the door, and said, 'Come, Mr. C—r, walk in; here's a good seat for you.' My courage failed me, and I sat down gentle as a lamb. I went home praying, and fell upon my knees, and never got up till I found salvation. I feel it in my soul to-day, brethren. Halleluiah!"

These words then swelled in melodious strains from many a joyous heart:

"By faith I view my Saviour dying,
On the tree! on the tree!
To every nation he is crying,
Look to me! look to me!
He bids the guilty now draw near,
Repent, believe, dismiss their fear-
Hark! hark! what precious words I hear,
Mercy's free! mercy's free!

"Did Christ, when I was sin pursuing,
Pity me, pity me?
And did he snatch my soul from ruin?
Can it be, can it be?
O, yes! he did salvation bring;
He is my prophet, priest, and king;
And now my happy soul can sing,
Mercy's free! mercy's free!"

Brother B—n said, "I embraced religion in Baltimore City, and the time which has since elapsed is the happiest period of my life. The grace of God has been sufficient for me in California. When I came to California, I knew but little about the doctrine of holiness, but I have here learned much, and have proved the virtue of Christ's blood in a *full* salvation from *all* sin. Glory be to God! I am the Lord's wholly, soul, body, and spirit. O, that I may be 'preserved blameless unto the coming of the Lord Jesus Christ.'"

We then sung:

"Refining fire, go through my heart;
Illuminate my soul;
Scatter thy life through every part,
And sanctify the whole.

"O, that it now from heaven might fall,
And all my sins consume;
Come, Holy Ghost, for thee I call;
Spirit of burning, come."

Then Brother H—t arose and said: "I am not ashamed of the Gospel of Christ, for I have proved it to be the power of God unto the salvation of my soul. I embraced religion eighteen years ago in the City of Philadelphia. But for religion, I believe I would now be in hell. God has been good to me. Religion is love."

Then the song:

"Religion is a glorious treasure,
Diffusion of the Saviour's love;

The Spirit's comfort without measure;
It joins our souls to those above:
It calms our fears, it soothes our sorrows,
It smooths our way o'er life's rough sea:
While endless ages are onward rolling,
This heavenly portion ours shall be."

Next arose Brother W—r, and said: "I have enjoyed religion for ten years. I have been a very unprofitable servant, but I am happy in God to-day. I have children in heaven; I expect to meet them there."

And then we sung of that happy home:

"A home in heaven, when our friends have fled,
To the cheerless gloom of the moldering dead:
We wait in hope on the promise given;
We will meet up there in our home in heaven."

Brother E—s next arose and said: "I was left a little orphan boy in Sweden. I soon after went to sea, and was discharged from the ship at a Swedish port, far from home. I was without friends, and without money, but I remembered that my mother used to tell me, that if ever I got into any trouble, to pray to God, and he would direct me. So I thought this is a time of trouble, and I went down to the sea-shore and prayed. I felt as though God heard my prayer. I rose from my knees, and walked along, cold and hungry, not knowing whither I went.

I had gone but a short distance, when my attention was arrested by the sight of some children going to school. I thought: 'O, I wish I could go to school, but I have no father and no mother; I am a poor homeless boy, with nobody to care for me,' and, weeping, I went and sat down on the steps of the school-house. The teacher came out and said: 'Little boy, where do you live?' 'I don't live anywhere,' said I; 'I have no home.' 'Where do your parents live?' 'I have none; my father and mother are both dead, and I have not a friend in the world.' 'Would you like to go to school?' 'O, yes,' I replied, 'I would be very glad to go to school, if I had clothes fit, and had anywhere to live.' 'What is your name?' said he. 'P. E—s.' 'Well, Peter, you shall live with me and go to school.' Thus, though not a Christian, God gave me an education in answer to prayer. I afterward became a sailor, and got very wicked. But soon after I came to California, in the early part of 1850, I heard Brother Taylor singing on the Plaza one Sunday afternoon, and I went up and listened to what he had to say. The truth took hold of my heart, and that week God converted my soul. I am happy to find so many friends here in California, lovers of Jesus. I believe, verily, that God will fetch me through."

And then our souls found utterance in Newton's good old hymn:

"Though troubles assail, and dangers affright,
Though friends should all fail, and foes all unite,
Yet one thing secures us, whatever betide,
The promise assures us, The Lord will provide."

Next Brother H—n said: "I obtained religion in Prince Edward's Island. Religion in California has been my polar star. I do feel that I have an interest in Christ's atoning blood. I thank God that on my first Sabbath in California I went to Church. That was a turning point with me."

Brother W—w said: "I have buried seven children in their infancy. They are all safe in heaven, where I expect to meet them. I have but one child living, and he gives me more care than the loss of all the rest. O pray that he may be converted."

Brother S—r, from New-Jersey, said: "I have proved the blessedness of religion by an experience of twenty-two years. I was converted in a love-feast, and I never attended one since without getting my soul blessed."

Brother O. said he had enjoyed religion for twenty-six years, and intended to travel "all the length of the celestial road."

We then sung:

"Even down to old age all my people shall prove
My sov'reign, eternal, unchangeable love;
And when hoary hairs shall their temples adorn,
Like lambs they shall still in my bosom be borne.

Brother H. B., from Baltimore City, said: "I was brought up without any religious instruction. I knew nothing about religion, had never even heard a prayer in the family till I was twenty-eight years old. I. D., my partner, was a Methodist class-leader, though I did not know it, for I knew nothing about the Methodists. One Saturday evening my partner said to me: 'Henry, if you go on in this way you will be lost. You ought to pray, and go to church, and seek religion.' Said I to him, 'What church do you go to, sir?' 'To Caroline street Church.' 'Well, sir, I'll go to-morrow,' said I. I went, and heard Thomas Seargent preach. The truth made a wonderful impression on my mind. I went the next day and bought a Bible and Jay's Prayers, and commenced reading and trying to reform; but I was completely miserable. For three days and nights I could not rest nor eat. I then went to a camp-meeting, the first I ever attended. This was the sixth day of August, 1833. That night Samuel Kepler preached, and invited all persons who wanted to seek religion to come into the altar. I immediately arose in the congregation and started in haste to the altar. When I got to the gate the gate-keeper said to me, 'We don't want any in here but mourners.' 'I don't know what you mean by mourners,' said I; 'but I want to seek salvation.' 'Come in, come in,' said he; 'you are the very man we want to see.' I kneeled down and cried mightily to God for mercy. A man said to me, 'Believe in the Lord Jesus Christ, and thou shalt be saved.' 'What did you say, sir?' said I. 'Believe in the Lord Jesus Christ, and thou shalt be saved.' 'I do believe in the Lord Jesus Christ,' said I; and at that very moment I did believe and experienced the salvation of my soul. Glory be to God. For many years I have enjoyed *full* salvation through the blood of Jesus."

"Praise God, from whom all blessings flow,
Praise him, all creatures here below,
Praise him above, ye heavenly host,
Praise Father, Son, and Holy Ghost."

CHAPTER XXXVII.

"YOU'VE KNOCKED ME ALL INTO A KINK."

ON Sunday, January 10, 1854, after preaching on the Plaza from the text, "If our heart condemn us, God is greater than our hearts, for he knoweth all things," a stranger spoke to me, saying:

"There is a man by the name of S., from B., lying at the point of death in that house, the third door from here," (pointing to the door.) He also intimated to me something of S.'s notorious character as a wicked man, and said he: "S. did not send for you; but his parents were religious, and perhaps you may do him some good."

I went in, and found him attended by four or five men, who appeared to receive me very kindly. He lay, pale and ghastly, evidently very near the grave. I said to him: "Friend S., do you suffer much pain?"

"No," replied he, very abruptly. I then turned away and exchanged a little conversation with his companions, and, in about five minutes, I approached him again, and, in the mildest and most hopeful manner I could, said:

"Friend S., do you not feel as though you might rally and recover?" hoping to gain access to his heart. He replied:

"When I want anybody to talk to me, I'll send for him."

"I have called in," said I, "as a friend, feeling the greatest sympathy for you, and am ready to do anything for your comfort in my power."

"I'd thank Mr. H.," said he, upbraiding the man whom he suspected of asking me in, "to attend to his own business." And then addressing me, he continued: "Before you came in here I had some peace, but you have knocked me all into a kink, and if you will just go away, I think I can die in peace."

He lived close to where I preached on the Plaza, and he had probably heard me preach a hundred times, and thus my presence, without the utterance of a word in regard to the condition of his soul, brought to his mind, doubtless, a thousand Gospel associations which seemed to throw him into unutterable tortures. His only peace depended on his banishing from his mind all thoughts of the past and future. Poor fellow! how sorry I felt for him. If the presence of a poor street preacher, clogged with mortality, "knocked him all into a kink," to use his own language, how could he bear the presence of holy angels, and of the great multitude of the redeemed in glory, were he admitted to heaven? How could he bear the presence of the awful God, whom he had insulted and defied all his life? How preposterous the idea of any man's being received into the kingdom of glory, without an education adapting him to heavenly enjoyments; a moral fitness for such a place. Heaven would be the most unbearable of all hells to such a man as poor S. He left the world "all in a kink," a few hours after I saw him, and eternal ages will not suffice to straighten him out. We have got to untangle all our kinks on this side of the river, or remain "all in a kink" forever. Let every man lay this to heart, "For there is no work, nor device, nor knowledge, nor wisdom, in the grave, whither thou goest." When we stand before the judgment-seat of Christ, it will be to answer "for the deeds done *in the body*." No record there of anything done by us *out of* the body, or subsequent to our leaving the body. The subject-matter for the final adjudication of the last day is all taken from the records of our probation on earth. Our case at death is at once entered in its place on the calendar of the Supreme Court of the Universe, and there it unalterably stands for the day of trial. Sinner, beware! "God out or Christ is a consuming fire."

CHAPTER XXXVIII.

HONORARY CHURCH MEMBERS.

At the close of a camp-meeting, held in Alameda, in May, 1854, twelve miles from this city, across the bay, at which fifteen persons professed justifying grace and united with the Church, I commenced a protracted meeting in the Bethel, which was continued a month. During the Bethel meeting upward of thirty professed to experience religion, and united with the Methodist Episcopal Church. When J. B. B., one of our young converts, gave his name as a candidate for Church membership, he said:

"Brethren, in almost all societies there are active and honorary members. It is never expected that honorary members should do anything in the Society. Now I have not come into the Church as an honorary member. I want you to put me down for an active member, for I want to do all I can in the cause of God. I have a burning love for souls, and I mean to do all in my power to bring them to Jesus.'"

He was very active as a young Christian in the Bethel. He is now a member in Powell-street. I baptized him and several other young converts in the bay, by immersion. Some of our ministers decline to administer baptism by that mode, believing it much better to convince the subject that "sprinkling or pouring" is better. But if a person has been brought up under Baptist influence, and taught from childhood that they ought to be immersed, I generally find it better, as I believe, to settle their conscientious scruples at once, by putting them under the water. At any rate, I feel bound to go by the Discipline.

"O, I'M SO 'SHAMED!"

During the progress of a protracted meeting in the Bethel, in July, 1854, I said to a sailor who seemed to be concerned, as I thought, about his soul, "Come, sir, come along, and kneel down at the altar." He, thinking that I was captain of the ship, and that my orders were not to be questioned, got right up and promptly obeyed the order.

After a while I went to him to give him a little instruction in regard to the work before him, when he said: "O captain, do let me get up, I feel so *'shamed;* I have nearly fainted two or three times. O, I am so 'shamed, I must go, I can't stay here. If you will let me go this time, I'll come back to Church again next Sunday. Do just let me off this time."

"Why, my dear sir," I replied, "if you get up and go out now, before all this congregation, they will look at you, and will think you are 'backing down' from what you have undertaken. You had better remain where you are till the meeting is out."

"O, I am so 'shamed," he responded.

He remained on his knees till the congregation was dismissed, but I could not get him to pray much. He left, and I saw him no more. Poor fellow, he would have a hard time of it if admitted into heaven in his sins and shame.

CHAPTER XXXIX.

THE SAILOR'S VISION ON LONG WHARF.

Two English seamen heard a sermon on Long Wharf, in the autumn of 1853, on the healing of the woman who had been sick twelve years, and "had suffered many things of many physicians, and had spent all that she had, and was nothing better, but rather grew worse," until she found Jesus, and touched the hem of his garment.

They became so distressed on account of their own wretched condition as sinners, that they went to the Bethel that night, and presented themselves as seekers of religion. Soon afterward, they experienced the healing virtue of the blood of Jesus in their own hearts, and became consistent, happy Christians.

One of them afterward, in relating his experience, said: "When I saw that poor old woman, on Long Wharf, press through the crowd, and touch the hem of the Saviour's garment, I couldn't help but cry, and I thought, O I wish I could go to him and touch his garment too, and be healed with the poor woman."

William B., a zealous young Christian in our

Bethel, speaking of what a miserable time he had while in the service of the devil, gave an account of his voyage to California in 1848. Said he:

"I shipped in the brig C. F., Captain P., from Baltimore. After we got out to sea, the captain flogged me regularly three times a day, all the way out, and called me by no other name than 'Son of a b—.' On one occasion he said to me, 'B., I believe whipping don't hurt you much, and now I am going to punish you.' He took me and tied me over the hawser-pipe, at the bows, where I was drenched with sea-water at every dip of the brig. I remained there in soak, without a bite to eat, for three days and nights. The captain also beat the cook till he jumped overboard, and then lowered a boat and beat him in the water, and took him up just in time to save his life. I was then a wild, drinking boy, nineteen years of age."

A sailor's life is a hard one at best, but poor B. seemed to have fared worse than usually falls to the lot of his kind. Flogging has been abolished in our navy, and is but seldom resorted to in our merchant ships at the present day.

CHAPTER XL.

"A SABBATH-DAY'S JOURNEY" IN SAN FRANCISCO.

I WILL take my reader back to Sunday, the twenty-seventh day of August, 1854. Not because of the peculiarities of that day particularly, for there were many of the same sort, but because I happen to find in my journal copious notes of that day, which bring all its scenes fresh to my memory. I will ask the reader's pardon before I start, for the apparent egotism of requesting his company through a whole Sabbath, to hear me sing, and preach, and walk, and talk; but then you must know that I only claim to be a poor sinner, saved by the great mercy of God, in Christ, taken up from the humble walks of life, and can never forget "the rock whence I was hewn, and the hole of the pit whence I was digged." Whatever good is wrought through me, the Lord doeth it. To him be all the glory. I commence my journey at half past eight in the morning in the U. S. Marine Hospital, and go through all the wards, and distribute tracts to one hundred and fifty patients. See how glad they are to get the tracts. There is a poor fellow almost gone. Tracts are of no avail to him now. He has doubtless stood many a gale, but, poor man, he is "stranding" at last. I will speak to him of Jesus, and have a word of prayer by his side. There's a man who has been confined two years by an abscess in his side. And there's a poor fellow who has been still longer confined by paralysis. He is unable to talk, but he can hear, and understand, and can read. I will tell him of the great Physician, who is famous for curing the "palsy," and give him a good tract. And now, at nine o'clock, the bell rings for preaching in the dining-room. I will sing while they are coming in.

"By faith I view my Saviour dying,
  On the tree! on the tree!
To every nation he is crying,
  Look to me! look to me!
He bids the guilty now draw near,
Repent, believe, dismiss their fear;
Hark! hark! what precious words I hear,
  Mercy's free! mercy's free!

"Did Christ, when I was sin pursuing,
  Pity me, pity me?
And did he snatch my soul from ruin?
  Can it be, can it be?
O yes, he did salvation bring;
He is my prophet, priest, and king;
And now my happy soul can sing,
  Mercy's free! mercy's free!

"This precious truth, ye *sailors*, hear it,
  Mercy's free! mercy's free!
Ye ministers of God, declare it,
  Mercy's free! mercy's free!
Visit the *sailor's* dark abode,
Proclaim to all the love of God,
And spread the glorious news abroad,
  Mercy's free! mercy's free!

"Long as I live, I'll still be crying,
  Mercy's free! mercy's free!
And this shall be my theme when dying,
  Mercy's free! mercy's free!
And when the vale of death I've pass'd,
When lodged above the *stormy blast*,
I'll sing while endless ages last,
  Mercy's free! mercy's free!"

"The lesson for the occasion may be found in the third chapter of the Gospel of St. John, from the first to the seventeenth verse.

"We will now sing that good Church hymn, commencing:

"'Of him who did salvation bring,' etc.

"Your mothers taught many of you to sing when you were little boys. Now, can you not all unite with me in singing the praise of God? Sing out just as you do sometimes when 'hauling ship into port.' Now, let us all kneel down and pray. Jesus is here this morning; he's looking at you now, 'and his ear is open to your cry.' O speak to him!" There were present about forty convalescent patients. The text may be found in the fourteenth and fifteenth verses of the third chapter of St. John: "As Moses lifted up the serpent in the wilderness, even so must the Son of man be lifted up, that whosoever believeth in him should not perish, but have everlasting life."

"I. The scene in the wilderness. The camp of Israel, containing nearly two millions of souls, more than six times exceeding in number the entire population of this state. The Arabian desert, as far almost as the eye can penetrate the distance, is covered with the tents of Jacob. What a vast 'camp-meeting' that was, and the presence of God was manifested there. But 'the people spake against Moses, and against God, Wherefore have you brought us up out of Egypt to die in the wilderness? For there is no bread, neither is there any water,' (that was a lie, upon their own admission,) 'and our souls loatheth this light bread.' 'And the Lord sent fiery serpents among the people, and they bit the people, and much people died.' You have been in large cities where the ravages of cholera threw the whole population into a panic, but throughout this vast encampment there was one universal wail of distress. The human family has an awful dread of snakes, especially when they crawl into our tents at night. But here a swarm of the most dreadful serpents we ever heard of, poured simultaneously from all directions into the camp. They ran like racers, and flew from tent to tent like angels of death. They were so quick in their movements, that there was no possibility of resisting them, and the poison they struck, as by electric shocks, into the life-blood of old and young, in every tent, was so deadly, that there was no remedy. This is the picture of the moral condition of the whole human family. Where is the man or woman, in this world, who has not been stung and poisoned by the old serpent that invaded the very garden of Eden? The effect, mortally, is not so sudden as in the camp of Israel, but none the less certainly fatal to soul and body. O the misery, and lamentation, and woe, occasioned in all our tents by that deadliest of all poisons, sin!

"II. What is the remedy? Moses, by the command of God, made a serpent of brass and put it on a pole. And as Moses lifted up the serpent in the wilderness, even so has the Son of man been lifted up, etc.

"III. What is the condition by which the healing virtue of the remedy may be applied to the subject? A proclamation was sounded throughout the camp of Israel, 'Look at the brazen serpent and live.' And from all the tents you see them coming forth, some walking, some crawling, some carried by their friends; and looking, they lived. So soon as a sufferer fixed his languid eye on the brazen serpent, his eye brightened, the poison in his veins was neutralized, and he sprang up in vigorous health. Some almost dead, their eyes already set in their sockets, yet when raised up by a friend, and their dim eyes fixed on the object, they immediately sprang into life. But there's an old skeptical, stubborn Jew, who is badly bitten, who says, 'What good will it do me to look at a piece of brass? Who ever heard of a piece of brass curing a snake-bite?' It is the power of God that is to cure you, my friend; but he requires you to *look*, as a necessary condition and test of your obedience. 'But,' says he, 'it will do me no good and it's no use to try. If God intends to cure me, he'll cure me anyhow. If he does not, I can't cure myself by looking at the brazen serpent.' Need I tell you, my hearers, that the stubborn old sinner died in his sins without remedy? So also a dying world is invited, by the glad tidings of salvation, to look to Christ, the uplifted Saviour," etc.

The poor, sick sailors listened with great attention.

I must now hasten to my appointment, at ten o'clock, on Davis-street Wharf. I will not trouble you to keep up with me. I'll run on, and sing up a crowd, and you can come along at your leisure. The chariot and the royal proclamation sung, see what a multitude of hungry souls, of almost every nation,

I have tc preach to! O, Redeemer, pity these lost sheep.

Our text reads as follows: "Daughter, be of good comfort: thy faith hath made thee whole; go in peace."

"The dreadful condition of this poor woman: sick for so many years: spent everything she had for medicine and doctors' bills: suffered many things by their experiments: got worse and worse. Such is your *moral* condition to-day.

"What was the poor woman to do? She had heard of Jesus. She saw from her window a crowd in the street, and inquired, 'What is that? what is that?' 'Jesus has come to town,' said one. 'Yes, that's the great Prophet of Nazareth passing along the street now.' 'O! if I could only get to him,' thought the sick woman, 'he would cure me of this dreadful plague, and would not charge me a penny.' But, poor creature, how could she? 'And yet,' thought she, 'this is probably my last chance. I may never see him again. I had as well die in the street, and be trodden down by the multitude, as to die in this miserable place. I'll go to Jesus. *I'll go*, or *perish* in the attempt!'

"Give back, gentlemen, and let this poor woman pass. Struggling under untold miseries of body and of heart, overcoming one difficulty after another, see her approaching the Saviour. O! if she can only get to him she will be cured, for he is full of compassion, and never turned one poor sinner away who came to him for relief. She crawls up behind him, almost fainting with exhaustion, and 'touches the border of his garment.' See her spring to her feet—active as a girl of sixteen. Now, though cured, her conscience is tender, she fears that she has intruded, and she would not offend Jesus for the world; but she makes a candid confession, right there in the street, in presence of the whole crowd, and Jesus says to her, with a smile betokening the infinite sympathy of his heart, 'Daughter, be of good comfort: thy faith hath made thee whole; go in peace!'

"Can she ever forget that day? Can I ever forget the day when Jesus spoke peace to my soul? Glory be to God! Jesus is here to-day. He is passing along the street now. Ye weary, heavy-laden sinners, he is speaking to you. Hark! he says, 'Come unto me, and I will give you rest.' Will you go to him? Will you press your way in spite of all opposition? Will you go now? Sing the Doxology. Preaching at the Bethel at eleven o'clock. *Now*, as soon as we can get there. Come along, I'll show you the way."

Text in the Bethel at eleven o'clock: "If any man will do his will, he shall know of the doctrine." Had a very gracious season. But if the reader will have patience to perform the "journey" of the day, I will not tax him with all its varied exercises.

At half past two o'clock in the afternoon I preached, in the hall of the Sons of Temperance, the funeral sermon of Robert Anderson, late of Mobile, Alabama. Text: "O death! where is thy sting? O grave! where is thy victory?" Mr. Anderson's family, consisting of his wife and three or four children, was present. His wife and daughter are Methodists. The old gentleman was nominally a Presbyterian. He, it is said, had been very wealthy, but had become by some reverses very poor. He informed me that for thirty years he had been a praying man, and had kept up prayers regularly in his family; "and yet," said he, "I never had religion. I believe I have been a sincere seeker of religion, but never knew anything of it but its forms." I spent hours with him at different times, before I could get him to look to *Christ alone* for salvation; but finally, some days before his death, his mourning was turned into joy. I believe he was genuinely converted, and that he conquered death, through the blood of Jesus, and has gone to a home in heaven. I accompanied the funeral procession to "Lone Mountain Cemetery," a distance of three or four miles. Five o'clock finds me back on the Plaza, attended by about one thousand hearers, to whom I preached a temperance sermon.

"My text on this occasion may be found recorded

in the Book of Chronicles of the 'Common Council' of this city. It is embodied in an ordinance passed by that honorable body last Monday night, the twenty-first instant, under the effect of a judgment rendered by one of our courts against the city, in favor of Mrs. Rosa Greenough, for ten thousand dollars. The said ordinance orders the payment of the said ten thousand dollars to the said Mrs. Greenough 'What was the ground of her claim against the city fathers?' She sued them for an indemnity for the loss of her husband, Robert Greenough, who fell through a hole in Bush-street, which caused his death. 'How did this hole happen to be in the street?' By the neglect of the city authorities to keep the street in order. 'What were the man's eyes for but to look out for holes in the street?' We have darkness as well as light, and when men *walk in darkness* they cannot see their danger. 'Why should the man be out in the dark?' That question is not relevant to the point. He *was* out in the dark, and returned to his waiting wife no more. He fell through the hole and perished. Had the city fathers done their duty, the hole had not been there, and Robert had not died at their expense. They confessed judgment, and paid the ten thousand dollars damages. Very good. If the man was worth that amount, and that is a very low price for a good husband, (though we can supply good ones here in California at a cheaper rate,) Mrs. Greenough had a right to the money.

"Well, on the very night this appropriation was made, a man by the name of Mahan got drunk and fell off Meigs's Wharf into the bay, and was drowned. 'How did Mr. Mahan come to his death?' He fell into a rum 'hole,' and perished in consequence of his fall. 'How came the "hole" there?' Through somebody's neglect? No; it was opened on purpose to catch men. 'Ah! do we have such holes in our streets?' Yea, verily. Not in Bush-street alone, but in every street in the city, and on almost every corner of every street. Are not these holes much more dangerous to life and limb than such holes as caused the death of poor Greenough? I believe Robert Greenough is the first man I remember to have heard of who lost his life in that way in this city. Who can tell how many hundreds of men, strong men, fathers of dependent families at home, and sons of affectionate mothers far away, have fallen into these rum holes and perished without hope? 'Their name is legion.' You have all seen the enormity of this evil in our midst. Does not a tremendous responsibility attach somewhere? Are not heavy damages due from some source? What is to pay? Ask the wife of H. S., whose husband was picked up in one of these 'holes' one dreadful stormy night, and was put by a policeman into a bunk in the station-house, not fifty yards from this spot, where he could be sheltered from the storm, and there, all alone, at the midnight hour, from the effects of his fall in these 'holes,' he died. He was once the head clerk of a large business house in B., a capable man, with an interesting wife and child waiting for his return.

"Ask the mother of Judge B., one of the brightest stars of the legal profession in this city. Many of you have hung with delight on the eloquence of his lips. But he walked in darkness, fell into these holes, and perished. He is gone forever. Who is to pay for all this? Do the city authorities, whose business it is to remove nuisances, stop dangerous holes in the streets, and protect the lives and property of their people, know that there are such dangerous holes in the city? Know it? How can they help but know it? Every five year old child in the city knows it. 'How is it that the city fathers seem to be ignorant of these things? Would it not be well to stir up their pure minds by way of remembrance?' Why, my dear sirs, what are you talking about? These holes are opened, and kept open, by their permission and authority. Their children are falling through these trap-doors of hell, into the burning pool, every day, and yet the fathers keep them open every day and Sunday, from the first day of January to the last day of December. O shocking! O consistency, thou art a jewel not to be found in the administration of civil government. Now, then, are not the city authorities in justice and equity responsible? If Mrs. Greenough could make them pay her ten thousand dollars for the loss of her husband, because of their neglect to stop a certain hole in Bush-street, has not Mrs. Mahan as good a right in equity to the same amount for the loss of her husband, because of the '*holes*' they *opened* in the streets? Why a man made them pay him eight hundred dollars for the loss of his horse, that fell through a hole in Pacific-street. 'Is not a man much better' than a horse? Here is Judge Shattuck, whom you all know, judge of the Superior Court in this city; I will submit the question to him without further argument, on the grounds of justice, equity, and consistency. Mrs. Mahan *vs.* the City of San Francisco. The plaintiff in this case, may it please your honor, sues for ten thousand dollars damages, for the loss of her husband, who perished last Monday night, by reason of certain dangerous 'holes,' opened in different parts of the city by the authority of the 'city fathers.' Now, gentlemen, we will wait the judge's decision. 'I am bound,' said the judge, 'in equity, to give judgment against the city, in favor of the plaintiff.'" "Good." A solemn appeal was then made to all Christians, patriots, and philanthropists, to oppose, by all honorable means, this sum of all evils, the rum traffic.

At half past seven P. M., I am again in the Bethel. Good congregation. Text: "Friend, how camest thou in hither, not having the wedding garment?" The occasion was one of great interest. Prayer-meeting after sermon. Singing lively and prayers fervent. I close my journey for the day in the "guest chamber" of a bridegroom and his blooming bride, whom I united in the bonds of holy matrimony. Wishing our reader companion pleasant dreams and refreshing sleep after a Sabbath-day's journey in San Francisco, I bid him good night.

CHAPTER XLI.

THE FARE HAS RISEN.

On Sunday, September 10, 1854, I preached at the corner of Davis and Jackson streets, to a large crowd, many of whom were miners, who had been disappointed in a passage home. The fare had been very high for many months, from $140 to $250 per passenger to New-York. But three steamers had advertised to leave on Friday, the 8th instant, and competition ran so high, that the fare was reduced to a very low figure. This caused a "rush from the mountains." Applications for passage were so numerous, and many were so desirous to get the fare reduced still lower, that the companies, the day before the sailing of the steamers, raised the fare to the former high rate. My text on this occasion was: "And the Lord shut him in." Noah and his family had wisely improved their opportunity, and secured their passage in the ark; but now the office was closed, no more passengers received, and the door was shut.

One of my illustrations ran thus: "Many of you wanted to go home. The fare to New-York was down to $20. That was a rare opportunity, but you delayed to embrace it. You said, 'Time enough yet. The fare will be lower still. By to-morrow we'll get tickets for a dollar and a quarter a piece.' And while discoursing on the probabilities of a 'more convenient season' for securing your tickets, one of your party inquired at the office to see how fast the price was coming down: 'What is the fare now, Mr. G.? 'I'll put you clear through to New-York in the steerage,' replied Mr. G., 'for $140.' 'Well, sir, I won't go in your boat,' said the miner; 'I'll wait for the next steamer.' 'Very good,' said Mr. G; 'whenever you get ready to buy your ticket, we'll be happy to sell you one.'

"The time has been, my friends, when you might very easily have secured a passage to heaven in the ark of salvation. You remember the many gracious opportunities you have had; but 'the fare has gone up.' It has risen in proportion to the multitude and magnitude of your sins. O how rapidly it has gone up since you came to California. You need never again talk of a convenient season to come. Your convenient season has passed long ago, never to return. The fare will never be any lower than it is to-day. But the office is open now, and your tickets are ready for you. We demand a renunciation of all your sins, and a consecration of your hearts and lives to God, and your belief in the Lord Jesus Christ.

Will you accede to these terms? 'Behold! now is the accepted time.' Engage your passage to-day. To-morrow the office may be closed, and the door shut, and then the deluge of retributive justice will sweep you away and drown you in perdition."

CHAPTER XLII.

A DUELIST'S FUNERAL.

I SUBMIT the following letter, as an introduction to my subject:

"Rev. MR. TAYLOR,—

"DEAR SIR,—Colonel Woodlief, a gentleman from Texas, with whom you probably had some acquaintance, was killed yesterday in a duel with Mr. Kewen. Previous to the duel in the morning he expressed a desire that, in case of his death, you should be requested to perform the appropriate ceremonies over his body. If you will be kind enough to do so, sir, you will confer a favor upon the many friends of Colonel Woodlief, and particularly upon his lady. The funeral will take place at two o'clock this afternoon, from the Tehama House.

"Very respectfully your obedient servant,

"RICHARD W. ALLEN.

"SAN FRANCISCO, *Nov.* 9, 1854."

Colonel Woodlief's untimely death was sincerely regretted by the large assembly of his friends who attended his funeral. It is not an easy task for a minister, in the presence of such an auditory and a weeping widow, to do justice to the cause of truth and the feelings of his hearers. I once heard a minister preach at the funeral of an alderman in this city, and, though the man was known to be a notorious drunkard, and it was believed that he had killed himself by hard drinking, he was held up by the minister, in the presence of the mayor, councilmen, and a vast assemblage of citizens, as a paragon of moral excellence. The impression was conveyed that he had without doubt been admitted to glory, because he was an honorable alderman of the City of San Francisco. My moral sensibilities were shocked. I would not unnecessarily hurt the feelings of bereaved friends. But thus to obliterate moral distinctions in character, and endorse such men, without repentance, as suitable subjects for the kingdom of heaven, gives the lie to God's holy word, and encourages sin. My fears for the effect of that sermon on the community were such that I was led, on the following Sabbath, to preach to a large audience on the Plaza from this text: "In hell he lifted up his eyes, being in torment."

On the occasion of Colonel Woodlief's funeral I said: "My dear friends, you are doubtless all acquainted with the person and character of Colonel Woodlief, and the melancholy circumstances of his death. He was, by birth, a fellow-Virginian with myself, and was always, I believe, regarded, by those who knew him, as a high-minded, honorable gentleman, and I exceedingly regret that I cannot add, a Christian. He was one of my regular hearers on the Plaza, and was often deeply affected by the word of truth. Some months ago, just after a sermon there one Sunday afternoon, I said to him: 'Colonel, allow me to introduce you to Captain M'Donald.' Taking him by the hand, the colonel said, 'I know the captain very well; we fought side by side on the fields of Mexico.' 'Ah, indeed! and did you know,' I replied, 'that the captain has embraced religion since he came to California?' 'O yes,' said he; 'I know that too. He told me all about it.' 'Well,' said I, 'do you see what a great change it has wrought in him?' 'Yes,' said he, 'I see it, I see it.' His eyes filled with tears, and his utterances were choked by strong emotion. When he could speak, he said: 'Don't talk to me on that subject, I cannot stand it.'

"That was a gracious moment for Colonel Woodlief. The Holy Spirit was touching the tender chords of his soul, and wooing him toward the cross of Jesus. O, how sorry I am to-day that he did not yield to its blessed influence, and become a Christian! Religion would have made him a happy and useful man; and we would have been spared the mournful duty we are called upon to perform to-day. For had he possessed the love of God in his heart, the probability is he would not have been challenged; and had he been, he would have acted under a higher 'code' than that adopted by chivalrous, though erring men. He would have exhibited a *moral heroism*, in standing for his duty to God, himself, his wife, and to society, that would have put to shame the *moral* coward that would engage him in mortal combat. O that he had obeyed the calls of God's Holy Spirit; then, had he died in the order of Providence, we would stand round his corpse with very different feelings. We could then, indeed, mix a sweet solace into the bitter cup of the weeping widow. Beware, my friends, of grieving the Holy Spirit! Seek, while you may, God's pardoning mercy. Place yourselves under his parental protection, as obedient children, that you may be saved from, or prepared for, the dangers and death incident to mortal life. Jesus Christ, your best friend, is waiting now at the door of your hearts for an answer. He is very desirous to save every one of you from your sins, and only asks your consent.

"Will you let him? I sincerely condole this dear woman in the deep sorrow of her very sad and sudden bereavement. Let us bear her, in the hands of earnest prayer and faith, to the mercy-seat, and commend her to the compassion of Jesus."

The occasion was one of great solemnity and sorrow. We urged the audience to use all their influence to put down the murderous "code," on whose "honorable" altar their friend had been dishonored and sacrificed.

CHAPTER XLIII.

RESTITUTION.

The law of restitution is one of God's immutable laws. The man who willfully wrongs his fellow-man cannot obtain the favor of God, until the spirit of this law is fulfilled in his heart; when, if it be within the range of possibility, he will give proof of it by actual restitution. This law applies as directly to theft of character as it does to theft of property. The man who, by detraction and slander, robs his neighbor of his good name, cannot obtain pardon from God, until by confession to the injured party, and reparation of the injury done, so far as it is in his power, he fulfills the demands of the law of restitution. Many persons, and not a few professors of religion, are engaged in the miserable business of peddling slang, wholesale and retail; and yet they wonder why they do not enjoy the blessed assurances of Divine favor, as in other years. The fact is, the Lord can as easily save old Lucifer from the pit, as save the slanderer, unless he submit to the claims of this immutable law of restitution.

The practical operation of this law was illustrated by the case of a sailor who embraced religion in the "Bethel," in San Francisco, March, 1854. He wrote a history of his awakening and conversion in verse, which was published soon after in the California Christian Advocate. For the illustration of the point in question, I will here insert an extract from a letter I received from him the day he sailed from that port:

"SHIP WESTWARD HO, lying in SAN FRANCISCO BAY, *March* 21, 1854.

"KIND FRIEND,—I am afraid I did not give you such a cordial reception as I ought this morning; or thank you enough for your kindness in bringing me those books. But you will pardon me, as your visit was unexpected; and I had just come out from a set of drunken sailors in the forecastle, where I had been to get my breakfast. So be kind enough to receive my earnest thanks now, that I have time to address myself to you; though my ideas are very much confused, as I am forced to write this in the forecastle among a set of sailors, who are still half drunk, and swearing and talking around me.

"I was very glad to see you this morning, as I was getting rather discontented with my situation, having heard a very bad name of the ship, but your visit drove my discontentedness away at once, and I now feel quite happy. I have been thinking of, and praying to the Lord all day, and so have been able tc keep evil thoughts out of my head.

"I will now give you a brief sketch of my life. I was born in Chester, England, and brought up in London. My friends are all religious. My father died when I was thirteen. I then went to live with my uncle, E. D., who is now, with his wife and family, living at New Town, Geelong, Australia. I got tired of my uncle's house, and he apprenticed me to a baker at Bracknell in Berkshire, England, in 1847. I served the baker about four months, when I robbed him and ran away. Since then I have been going to sea. I came here in the 'Flying Dutchman,' in October last, and it is to make restitution to the baker that I am now going to England, by way of Calcutta and the States, not trusting to that act for my salvation, but to show my friends that my repentance is sincere.

"Since I have been at sea I have given myself up to all sorts of wickedness, and I believe I have not been more than half a dozen times to a place of worship during the whole term of my sea life till I came to California. Since then I have heard you preach several times. O, sir! if my friends in England only knew what a change has come over me, how happy they would be. I know they are always praying for me.

"I often think of the text, 'Be sure your sin will find you out.' It has found me out often on the deep. Perhaps you noticed my teeth being broken. That was done by a fall from aloft when I was in the Mediterranean once. If I had died then I should have gone to hell. And several other times have I narrowly escaped death, because God kept me safe for this hour. On Sunday last some old acquaintances were trying to persuade me to cheat the boarding master, and go with one of them in a sloop for more than twice the wages I am getting here, but I refused and went to the Bethel, not knowing that these things were going to happen. At another time I should have gone with them in a minute. I can see the Lord's hand in it all."

Here we see a young man, under the promptings of the law of restitution, which "the Holy Spirit writes on truly awakened hearts," leaving the land of gold, where he desired to stay, and where he had a fair opportunity to make money, to circumnavigate the globe for the purpose of restoring what he had taken when a boy. God takes "the *will* for the deed," *only* when the deed is *impossible*.

CHAPTER XLIV.

"SHANGHAEING" THE SAILORS.

The humble, but just claims of the men of the sea, upon the consideration and sympathizing regards of the American people, have often been presented and urged, and as often disregarded and rejected, until within the last thirty years, since which time they have been, in part, acknowledged and honored. The history of the sailor, his isolation from domestic society and the refinements and luxuries of home, his spirit of adventure, courage, patience, toils, sufferings by starvation, cold, shipwreck, confinement in foreign hospitals, adventures among savages and cannibals, his imprisonments and slow tortures, his death by the violence of war and piracy, by the violence of the hurricane that sweeps the ocean, and by the more dreadful tortures of wasting famine, has been written in detached fragments on every page of the history of commercial nations, and especially of our own country. Strike out the history of the sailor, and American history would be what American commerce would be but for the presence of the living sailor. Seamen are our great explorers and discoverers, and the bones and sinews of our commerce; and they, more than any other instrumentalities which can readily be employed, can hinder or assist the dissemination of enlightened civilization and Christianity throughout the world.

The complaint has often been made that the influence of the seamen of Christian nations upon the heathen, was one of the greatest barriers to the successful preaching of the Gospel to heathen nations that the missionary has had to encounter. The principles and facts underlying this complaint, or the grounds on which it is based, will prove that seamen, if Christianized themselves, will be the most efficient auxiliaries in spreading the Gospel in heathen lands the missionary can employ. Every enlightened, active Christian sailor is himself a missionary by direct appointment of Providence. We cannot dispense with our regular missionaries to heathen lands, and they ought to be multiplied a hundredfold. They are our foreign generals in the King's army, but we need private soldiers as well. Why not enlist the sailors? Seamen are as necessary in moral exploration and discovery in distant lands now, as they have been in the past in physical exploration and discovery. They have many and great advantages in this work over the regular missionary.

First. In regard to distance. What trouble and

expense the missionary endures in overcoming this difficulty; to the sailor no part of the globe is remote, and his expense is just nothing.

Second. In regard to the facilities of communication after the field is reached. The missionary must spend years to acquire and speak the language, and then his access to the people is embarrassed by numerous obstructions. His themes of conversation are new and undesirable. The very fact that he is a missionary, come to teach a people who consider their own religion the very best in the world, excites suspicion, jealousy, and prejudice. To overcome this difficulty, it has often been found advisable for the missionary to connect with his care for their souls the treatment of their bodies "as the doctor."

But the sailor gains access to their hearts at once through the channels of their dayly thoughts and feelings, commerce, trade, self-interest. They have no prejudices against the sailor, such as bar their hearts against the missionary. He is but a common man like one of themselves, and his familiar intercourse with them enables him to pick up and use their colloquial language with wonderful facility.

Third. In regard to priority of time. Preëmption in new countries insures a great advantage, morally as well as physically. The sailor makes his mark in heathen countries a hundred years in advance of the call for a missionary meeting to devise ways and means for the appointment and support of a missionary in that distant field. When the man of God reaches his distant field, he not only finds the sailor there and perfectly at home, but, also, that he has been received by the natives for a century past as a *bona fide* representative of Christianity. The heathen, knowing nothing of the distinction between professing and non-professing Christians, set down in their account every man belonging to Christian nations as a Christian, and fix in their minds the moral standard of Christianity accordingly. When the missionary attempts to explain, and draw the line of distinction between himself and his illustrious predecessor, the sailor, the heathen will reply, "Your people are all sharp; they know what is good. If your religion is as good as you say it is, why do they not all embrace it?"

These various facts might be amply proved and illustrated from the history of the heathen nations of Asia and Japan, and the islands of the Pacific in general; but are they not patent to all? Even in South America, the familiar resort of seamen for two hundred and fifty years past, the American Protestant Churches have two missionaries, one in Buenos Ayres and one in Valparaiso, on the opposite side of the continent. I have briefly sketched these facts to remind the American people again of their obligations to the sailor, and to remind the Church of the important fact, that the conversion of seamen is of the first moment, if not absolutely necessary to the Christianization of heathen nations.

"But," says one, "we can more easily convert the heathen than the sailors."

When Jesus wanted missionaries to bear the title of "Sons of thunder," he selected from among the sailors of the Sea of Galilee, and had no difficulty in getting them converted. I believe he wants many such now, and think that if the Church will do her part on behalf of seamen, the Saviour will have no difficulty in having them reconciled to God, and qualified for their work. Poor sailors! what advantages have they had? You have had pious parents, the restraints of the domestic circle, and of family religion, good maternal counsels, Sabbath-school instructions, and preaching every Sunday since you were a little boy; but the mass of seamen have none of these. The little orphan boy was put aboard ship at the age of five years, and educated in the forecastle, under the tuition of the regularly graduated tars of the old school. His character bears not one molding touch of a mother's prayers or counsels, nor the refinements of home circles. The voice of prayer he never heard, except in the storm that wrecked the ship, but he heard the same praying ones swearing again, so soon as the storm abated. He never was at preaching since he was born, and the warm gushings of Christian sympathy never came in reach of his heart; and yet no heart is more susceptible of generous emotion, and more impressible by sincere sympathy than his. Those who seek to destroy him, well know how to take advantage of these facts. What class of men so deserving, have been so much neglected in the past? True, government has built them hospitals, and the Church, bethels and homes; but much yet remains to be done; and the extent of the abuse they have suffered in home and foreign ports, by the "landshark" fraternity, a system of abuse familiarly known in California under the title of "Shanghaeing," (pronounced Shanghiing,) has never been learned by the mass of our people, a "mystery of iniquity," the enormity of which the light of eternity only can reveal.

The following lecture on "Shanghaeing the sailors," was delivered on the Plaza to an attentive audience, in September, 1855, and I here give it as nearly verbatim as I can copy it from the records of memory. After making some remarks on the character and condition of seamen, I said:

"Gentlemen, the system of Shanghaeing, to which I invite attention, is almost as ancient as the commerce of nations; but the term Shanghaeing is a modern, California name, the origin of which we will give you in due time. I say system of Shanghaeing, because it embraces a combination of laws and forces, employed by a combination of men for the accomplishment of a specific end, namely, to reduce to a state of perfect vassalage and voluntary serfdom, the millions of men who 'go down into the sea in ships,' and to gather all the fruit of their toils. The secret motto of the system is, 'get all the sailors' money, honestly if most convenient, but get it.'

"A single Shanghaeing fraternity (and we have twenty-three of them along our water front) embraces, 1. A sailor landlord, alias 'landshark,' alias 'Shanghaer.' 2. A drayman. 3. A 'longshoreman.' 4. A sailor lawyer. 5. A shipping-master. The sailor landlord keeps a sailor boarding-house, bar, etc. The longshoreman mans, with a pair of oars, a Whitehall boat. The sailor lawyer prosecutes suits against captains and owners of vessels, and otherwise collects seamen's wages, damages for maltreatment, etc. The shipping-master provides crews for the ships as they 'clear,' by contract with the ship-master, for five dollars per head. The captain of the vessel ships none of his men directly; they must all come to him through the shipping office, where the shipping articles are kept for signature. The whole contract for the voyage with the crew, is made by the shipping-master, who is to see them all aboard at the hour for sailing, and the captain has nothing to do with them till he gives the order for sailing. The advance wages are paid, not in money, lest the sailor should spend it, and then refuse to go to sea, but by a check on the shipping-office, to be paid three days after the ship sails. The ship-master, it will be seen, is not a party in the Shanghaeing business, and the shipping-master is a party from necessity rather than from design, as we will show, and may, nevertheless, be an honorable man. The lawyer may, by possibility, be an honorable man, but he will bear watching.

"We will now show you the practical working of the system. A ship is 'telegraphed,' and the 'longshoreman' is ready with his boat. He is in the stream, and listening for the command, 'Let go the anchor.' Immediately he is on deck, and perfectly delighted to meet his poor brother seamen from a long voyage.

"'How are you, my good fellows? I'm glad to see you! You've got to the right port at last. The most glorious country in the world; a regular God-send for poor sailors! A crew came in last week, and left their ship, as they all do here, and now every man of them is getting a hundred dollars a month to stay ashore. If you'll all come along with me and put up at our house, I have chances waiting, and you shall have work ashore at once, and wages that are wages.'

"Thus he decoys the entire crew. Sometimes he takes them right away in defiance of the captain. On one occasion a captain ran out, with a small deringer in his hand, and the longshoreman said,

'Captain, what are you going to do with that thing?' 'If you interfere with my men,' replied the captain, 'I'll put a ball through you!' The boatman, pulling out one of Colt's large revolvers, said, 'Here, captain, take this; that little thing is no account.' Then, turning on his heel, he said to the crew, 'Come on, boys. Pass down your "donnage" here into my boat. I'll take care of you.' They all went in spite of the captain's threats. But the usual method is to make an appointment with the crew to have the boat 'alongside' at a certain watch in the night. The boys all get ready at the hour appointed. Their faithful friend, who has promised to emancipate them from the horrors of a sailor's life, is on hand, and bears them to the wharf. The drayman is waiting on the wharf with his dray, on which their 'donnage' is transferred from the boat, and now all hands march up together to the 'home.' The landlord meets them with a hearty 'shake hands.' 'Welcome to my house, my good fellows! You've had a hard time of it, I know. I am prepared to sympathize with you, for I am a regular old salt myself. I'm glad you've come to this glorious country! Walk in! walk in! and make yourselves at home. Everything in this world you want you shall have while you stay with me.'

"Now all hands must 'treat.' The landlord treats,

the longshoreman treats, the drayman and each sailor treats, and by this time the whole crew is drunk, unless, perchance, there may be one who has 'signed the pledge.' The longshoreman and drayman now demand their money, which the landlord pays—ten dollars a head for decoying and bringing ashore, and five dollars per head to the drayman—and charges to the sailors' account; and all the drinks, real and imaginary, are also set to their account.

"The lawyer now comes in, and, through the landlord, gets all their claims against the ship, which he collects, either by a compromise or a suit at law, as may best suit the convenience of the captain, half of which he keeps for his trouble, and pays the other half over to the landlord on behalf of the sailors.

"Next comes the shipping-master, who says, 'I want twenty men for the ship Water Witch, by five o'clock this afternoon.' 'Very good,' says the landlord, 'you shall have them.' And often the very crew that came in the morning, are shipped before they recover from the first 'drunk.' When they wake up from their golden dreams, they find themselves at sea, in a strange ship, minus their 'back' and 'advance wages,' and most of their clothing. A part of their clothing they find in their chests, and a bottle of whisky to keep to sober on, and to remember their friend, the landlord, by.

"When the sailor's bill at the boarding-house runs

up to cover the 'advance,' the landlord says to him, 'Jack, you must ship.' 'I won't do it,' says Jack. 'You shall do it; you owe me a hundred dollars, and you must either pay it to-day, or go to sea in the ship Challenge.' 'O, I don't want to go to sea yet,' says Jack. 'O, well, never mind,' says the landlord, 'you're a clêver fellow, and you may stay at my house as long as you please, and pay me when you get ready. Come, let's take a drink.' Jack, very glad to be on so good terms with the landlord, walks up to the bar, and drinks to the health of his master. In ten minutes he is as insensible as a log. When he recovers from his mysterious sleep, he is out of sight of land. He is awaked by the stern command, 'Wake up here, and go to work.' The poor fellow, rubbing his eyes, inquires, 'What ship is this? Whither bound?' 'To Hong Kong.' 'How did I get here?' 'Why, you shipped, sir,' says the master. 'I never shipped in this ship.' 'Yes, you did, sir, and you must go to work without any more grumbling,' replies the captain sternly. 'I want to see the articles,' says the sailor. 'Well, sir, here they are; what is your name?' says the captain. 'My name is John Waters.' 'There it is written on the articles in two places, once by the landlord, and once by the shipping-master.' 'I never signed those articles,' replies John. 'No,' replies the captain, 'you were too drunk to write your name, but there's your mark.'

John puts his hand on his head and studies a moment, and says: 'I want my advance before I go to work. How much was I to get?' 'One hundred and twenty-five dollars for the run, paid in advance,' replies the captain. 'And here's your account from the shipping office; your bill with the boarding master took one hundred dollars, leaving twenty-five dollars, which he handed to me to give to you when you got sober.'

"John takes the twenty-five dollars and goes to work. But you ask, 'What did the landshark give to the sailor to take away his senses so suddenly?' It was a compound of whisky, brandy, gin, and opium, which, if a man drinks, he sinks into the Lethean stream for a dozen of hours. In days past, when seamen were scarce in this port, very many landsmen, as well as seamen, were thus drugged and shipped. On one occasion a shoemaker stepped to the bar to take a 'drap,' and waked up the next day at sea, and did not get back to his business for nine months. A brick-mason, as I was credibly informed, was thus shipped in the ship Hurricane. Again, a drayman left his dray in the street, and went in to take a 'nip,' and saw his dray no more. I was told that a Spaniard, with his long spurs on, was thus shipped on the clipper Contest, Captain Brewster. It happened, however, that they had not given him quite enough, and by the time they got him aboard, he recovered and 'showed fight,' whereupon the shark knocked him down. But Captain Brewster, a humane gentleman, would not suffer such cruelty aboard his ship, nor take the Spaniard against his will. So the long-spurred 'hombre' vamosed.

"A man boasted that, having stabbed a fellow, he had escaped a term in the state prison by drugging and shipping his victim before the trial came on. How many homeward-bound miners have been thus drugged, and robbed, and shipped, eternity will reveal.

"Again, there are some men in California who will not drink rum, and the Shanghaer cannot dispense with the services of such, and the question is, How will he get hold of them? Well, sirs, they have what is called the 'Shanghae cigar,' which is thoroughly impregnated with opium and other poisons. The smoking of one is equal to a dose of chloroform, with more lasting effects. I will illustrate the practical importance of this cigar by a single case.

"A landlord, lacking a man to make up a crew, met a German glazier on Long Wharf, with a pack of glass on his back, and said to him: 'Hie, my good fellow, don't you want a job?' 'Yes, sir.' 'I want you,' said the shark, 'to put some glass in the stern of that ship,' pointing to a ship in the stream. 'Jump into my boat here, and I'll take you aboard.' So off they went. As the German sat in the stern of the boat, much pleased with the prospect of a good job, the shark said to him, 'Will you have a cigar, sir?' 'Yes.' So the glazier sat and puffed away, as he used to do in his 'Fader-land,' but before they reached the ship he tumbled over in the bottom of the boat. The shark threw his pack of glass into the bay, and running 'alongside' hailed, 'On deck there, lower away and haul up this man.' A rope was lashed round him, and he was hauled up. The shark ran into the captain's office, saying, 'Captain, I've got you a first-rate sailor here. He's a little boozy to-day, but he'll be all right by to-morrow,' and got his advance. The poor German waked up at sea with a longer job than he had engaged for, and the worst of the business is, he must not only work for nothing, but be kicked and cuffed through the whole voyage for having the presumption to impose himself on the ship as an 'able seaman,' when he knew nothing about the business. These are the principal modes of 'drugging,' but they employ various other modes of 'Shanghaeing,' so that it is almost impossible for a man in any kind of communication with them to escape.

"A sailor, who was well acquainted with their 'arts,' boasted that they could not Shanghae him. One day, a landlord said to him, 'Tom, the clipper ship —— has made up her crew, and is ready for sea;

I am just now going to see her off. There are some of your acquaintances aboard, Bill Evens and Jim Jacobs; wouldn't you like to go aboard with me and see them before they leave? We'll be back in a few minutes.' 'Yes,' replied Tom, 'I would like to see the boys. I have business with Bill.' 'Very well, jump into my boat.' So off they pulled. On deck, Tom ran into the forecastle to see the boys, and the shark ran into the captain's office, saying, 'Captain, I have brought you a splendid seaman, the best man in port; and that makes up the complement. Here's his name on the articles.' So he delivers the papers to the captain and leaves. In a few minutes Tom came on deck to go ashore, and lo! the boat was gone. He had nothing to do but to obey orders and go to work. Thus to drown men's souls in rum, to poison, enervate, and destroy their bodies, and rob them of all their hard earnings, and leave their widowed mothers, wives, and children, who are dependent upon them, to beg or starve, is perfect sport for the 'landshark.' The great man-eater of the deep is satisfied to get the stray carcass of a sailor occasionally, but these dry land monsters must have soul, body, and estate of all the sailors, if possible. You ask, 'Why do not the sufferers have the fellows arrested, and brought to justice?' Because, 1. A man neither likes to confess that he was drunk, nor that he was so silly as to be duped and drugged. 2. He lost his senses so suddenly, and has been absent on his voyage so long, that he cannot think of one witness by which he can prove anything. 3. The whole fraternity is so powerful that the peril of an attack is more to be dreaded than a Shanghaeing.

"Arrests, however, have been frequently made here, as you are all aware, and sometimes justice is, in part, dealt out to them, but it is very hard to get evidence to convict them. You say again, 'Surely these California landsharks must be the worst in the world?' I know not, but I have heard of some very bad ones in all our large ports. Captain E. told me but a few days since, of the mate of an English ship which came to New-Orleans, who was drugged, and the next day found himself at sea in a strange American ship, shipped as a common sailor before the mast. Another case he gave of a man who was put aboard, it was supposed, 'dead drunk,' and his 'advance wages' drawn, but the next morning, when the captain tried to wake his man up, he found that he was dead, and had been so for a day or two. But you inquire again, 'Why do the sailors put themselves into the power of these fellows, and allow themselves to be so imposed upon?'

"By the attractive power, on the sailor, of false sympathy, promises of money-making, liquor, old acquaintanceships, bad women, etc., he is induced to desert his ship, and go to the house of his *dear friend*, the sailor landlord. According to law, as a deserter he can be arrested, and sent back to his ship, and made to perform the rest of the voyage for which he shipped without wages. So there the landlord 'gets him.' 'None of your cutting up about me. I'll tell your master where you are, and have you back on that ship before you can say "Jack Robinson."' The crest-fallen sailor 'gives in,' and is as humble as a whipped dog. Again, the landlord says to the shipping-master, 'You must ship your men from our houses. If you don't, we won't let you have a man when you want them. And you are to give us no trouble about those bills against the sailors.' The shipping-master is dependent, and must work into the hands of the 'sharks,' or be cut out of business. Now, then, if a decent sailor has independence of character enough to resist all the other snares, and selects a good boarding-house, when he goes to the shipping-office to ship, the shipping-master says to him, 'Where do you board, sir?' 'Up town, sir, at Widow ——'s.' 'We don't need men now; you can leave, sir,' says the shipping-master.

"The poor fellows cannot ship except through the shipping-office; and they won't have him there because he don't board in the right place. Now there are exceptions to the rules of the trade we have exhibited, and there are among boarding-masters some pretty decent men; but we have here revealed, we believe, truthfully, as the result of long personal observation, and good authority, the general workings of the system of Shanghaeing.

"This system, the same in principle in all large ports, varies in its practical operations according to local circumstances. The term Shanghaeing is, as we remarked in the commencement, of Californian origin, and was introduced in this way:

"A few years ago, as many of you remember, it was very difficult to make up a crew in this port, especially for any place from which they could not get a ready passage back to this land of gold. Crews could be made up for Oregon, Washington Territory, the Islands, and the ports of South America; for from any of these places they could readily return. Even from Canton, they could stand a pretty good chance of a direct 'run' back; but from 'Shanghae, in China,' there were seldom ever any ships returning to California. To get back, therefore, from Shanghae, they must make the voyage round the world. That was getting quite too far away from the 'placers' of our mountains. Hence to get 'crews' for Shanghae, except enough of 'Lascars' to get the ship to sea, they depended, almost exclusively, on drugging the men. Crews for Shanghae were, therefore, said to be 'Shanghaed;' and the term came into general use, to represent this whole system of drugging, extortion, and cruelty.

"Now in the light of all these facts, especially the desertion of all the crews immediately on the arrival of the vessels, you may see how difficult it is for a seamen's chaplain to gain an extensive influence over the mass of seamen in this city. In any other port, the preacher can board the vessels as they arrive; make the acquaintance of the whole crew, master and men; invite them to his house, and to his Bethel; and thus gain an influence over them for good. But here, when we board a ship, we find none who came in her but the captain; and though he, as a gentleman, will treat you kindly, still he is mad, and complaining of the landlords, and lawyers, and sailors, and the port; and you cannot get in the neighborhood of his heart, with any kind of *California moral* influence. He don't believe there is anything good about it.

"The seamen have gone under the dominion of the Shanghaeing fraternity, and cannot be reached, only as we 'take them on the wing,' by preaching in the streets. In this way, through the mercy of the Lord, we have seen many of them brought to God, and happily converted; and we have, under all these discouragements, maintained a self-sustaining Bethel for seamen, in this port, for four years.

"In conclusion, the question arises, 'What can be done to remedy these evils?'

"*First.* We want a good 'Seaman's Home,' under the direction of a good board of trustees. Such an institution not only provides for the shipwrecked and destitute mariner, but furnishes a good 'home' for all well-disposed seamen, where they can enjoy the elevating influences of good social society, bow at the family altar every night and morning, and be amply protected against the sharks. A good home, properly conducted, will so compete with the common sailor boarding-houses, as to hold in check their diabolic plans and purposes, and cause them to imitate the 'Home,' in order to retain custom. In short, it works an outward reformation among the 'sharks,' and insures pretty good treatment to sailors generally. The 'Home' is the place where shippers and captains of vessels, who do not wish to risk their ships and cargoes in the hands of drunken crews, can always go for sober, reliable, 'able seamen,' to man their ships; and there a pious shipping-master can successfully compete with such as would work into the hands of the sharks. Such an institution I have tried hard to establish in this port; but, owing to an extraordinary train of reverses which, during the year past, have befallen our city, my plans have been frustrated, greatly to my sorrow.*

* A society has been recently organized in this city, for the purpose of establishing a good Seaman's Home in this port. I sincerely hope they will succeed beyond their most sanguine expectations. September 25, 1856.

"*Second.* We want, in my opinion, a reform in the shipping laws, and practice, in regard to advance wages. The 'advance' was designed, no doubt, to work for the sailor's benefit by giving him, if poor, the means of supplying himself with clothing for the voyage. But, receiving his advance in the form of a check on the shipping office, as we have shown, to be paid after he has gone to sea, lest he should use the money and refuse to go, he is obliged to go to somebody who will trust him to get his check discounted. He generally has no friend who has money but his *dear* friend, the landshark, who takes the check, and, if he gives him any money at all, it is the remnant left after an extortionary discount has been taken off the face of the check. In most cases, I should say, the advance is a dead loss to the sailor, and, in very many cases, a serious injury; for he often has to take it all out in bad whisky. If the advance wages were cut off, it would be necessary for shippers and captains of vessels to lay in a little store of clothing for the crew, to meet emergences at sea, and also to see that their families were provided for, so far as they might suffer by the want of the advance. Then, if necessary to have shipping-offices and shipping-masters, as now, let also a copy of the articles be kept aboard in the captain's office, so that good seamen can go directly to the master of the ship and engage to sail with him, just as any other class of men do in their respective departments of business. Why should every honest seamen be subjected to the suspicious routine of the present system?

"*Third.* Let landsmen, and especially Christians, male and female, extend the hand of friendship to men of the sea, and manifest true Christian sympathy toward them. Let them labor as diligently for their physical improvement and the salvation of their souls as do the landsharks for their ruin. Much patience will be required for this work; for the sailor, so often wronged ashore, is very suspicious.

"An old 'tar' said once, 'There's a merchant who respects the Bible. The Bible says, take the strangers in. That's the commandment he keeps; for he took me in on a pea jacket.' The poor stranger from the sea has been 'taken in' so often ashore, that he, with too much truth, utters the complaint of exiled David: 'I looked on my right hand, but refuge failed me; no man would know me, no man cared for my soul.'

"We want good bethels for seamen; but we want more especially an earnest, patient manifestation of *personal* Christian sympathy shown to all seamen while in port. Convince a sailor that you are his true friend, and he is the most confiding, teachable man in the world. You can, by the grace of God, lead him to the cross of Jesus, and then to heaven."

At the close of the address I sung the "Dying Sailor's Lament:"

"The frown of the night-storm had scarcely blown by,
And the ocean was still in its roar;
The wind had not ceased from disturbing the sky,
When I ventured to walk on the shore.

"I look'd on the sea, and a wreck had been toss'd
On the breakers that roll'd from beneath,
And bodies, still throbbing, were wash'd on the coast,
And lay group'd in the stillness of death.

"I sought, from among the pale corpses around,
For some symptoms of life, but in vain;
When I heard, in the distance, an indistinct sound
Of a voice, that seem'd utter'd in pain:

"'Farewell, giddy world,' it exclaim'd, with a sigh,
'Disregarded and slighted by thee;
For my country I've fought, for my country I die,
But my country has cared not for me.

"'For thee, native land, my life I have spent,
And have spill'd my heart's blood in thy wars,
And yet, though your missions so far have been sent,
You've neglected the souls of your tars.

'"We were left on the brink of destruction to sleep,
And no voice hath aroused us away;
No arm was extended to collect the poor sheep
That had wander'd so sadly astray.

"'And now I must go to the doom that I dread,
Through ages that ever must roll,
With a life of iniquity heap'd on my head,
Yet there's "no man hath cared for my soul."

"He ceased, and I sought him among the pale dead,
While he yet had the hour to repent,
When a heart-rending groan, that yet thrills through my head,
Was the close of his hopeless lament.

"On the cold shore extended I found him at last,
But his spirit had ceased to be there;
His brow was still frowning, his hands were still clasp'd,
And he look'd the mute form of despair."

CHAPTER XLV.

JAMES KING, OF WM., AND HIS MORAL PLATFORM.

After all that has been said and written about James King, of Wm., very many of his best friends are anxious for further information in regard to his religious faith and prospects.

Again, the deep and almost universal sympathy and excitement, not only in this city, but throughout the state, produced by the assassination of Mr. King, is believed to be not mainly owing to personal attachment, great as it may justly have been, but to the fact that he was the people's exponent of many vital principles and questions of reform in California. It, therefore, becomes our duty to define as nearly as we can, the moral aspect and bearings of this mighty tide of public feeling, and voluminous expression of public sentiment; and to inquire how they may best subserve the great ends of moral reform in this state.

It is not my purpose here to repeat my views in regard to the organization and operations of the Vigilance Committee, nor to reiterate what has been so truly and so eloquently said in the public journals, of the excellences and usefulness of Mr. King, only so far as may be necessary to state and to illustrate our positions.

I wish first to call attention to Mr. King's moral platform, to which the thousands, who vie with each other in honoring his memory, have tacitly committed themselves; and, secondly, to his religious faith and prospects. Among the leading positions of his platform are the following:

1. Anti-dueling.

Dueling, that bloody code which demands the violation of every other law, the surrender of every ennobling social and moral principle, and the sacrifice of life. The man who submits to its claims is entitled, as a reward for his loyalty, to have his heart perforated with a discharge of cold lead, his brow encircled with a wreath of clotted blood, and to have inscribed on his tomb-stone, if he happens to have a friend left with respect enough for him to erect one to his memory, "Fell in a duel." None who gaze upon it envy his "honors," and many exclaim, "He died as the fool dieth." Strange as it may seem, this horrible code has obtained, to an alarming extent, in almost every state of our glorious confederacy. The State of Illinois is an exception. In a duel fought there, in the early settlement of that state, one of the combatants was killed; the other was tried, condemned, and executed for murder. That ended the "honorable code" in the State of Illinois. That convention of California pioneers, which framed our State Constitution, wisely ignored this barbarous code, and pronounced it a sin against the state to resort to it, by which a man would forever forfeit the privilege of holding any office in the gift of the people. Had we honored that law, it would have honored us, and saved many precious lives. But it has been utterly disregarded.

The case of Mr. Gilbert, a leading editor in our city, is fresh in the memory of many of you; shot down by a member of the Legislature, who resumed his seat unrebuked, and was soon afterward promoted to a high official station, which he has held nearly ever since.

Colonel Woodlief, poor fellow! He used to hear me preach on the Plaza, and I have seen him weep under the appeals of truth like a child. O, had he yielded to the gracious impulses of the Divine Spirit upon his heart, he might have been a living, happy man to-day; but he stood up as a target to be shot at, and was suddenly launched into eternity. Their name is "legion" who have engaged in this miserable business.

The King platform repudiates wholly this barbarous code. James King had the moral courage to expose the moral cowardice of such as felt it necessary to resort to such means to manifest their courage and vindicate their honor. When challenged, it was enough for him to say, that he sustained relations to God he was not at liberty to violate, and would not behave so dastardly as to rush unbidden into the presence of God, and leave his wife and helpless children to mourn at once his folly and his loss. Does not such a refusal reflect more honor upon a man than a hundred duels could possibly do?

I will simply enumerate other prominent positions advocated by Mr. King, as,

2. Anti-gambling.

3. To drive out of the city all houses of prostitution.

4. To expose "corruption in high places."

5. To purify the ballot-box, and promote *none but honest men to office.*

6. To furnish employment for the industrious poor who seek a home in our new country.

7. To promote public schools and educate the masses.

We would respectfully submit an amendment to Mr. King's position on this subject, namely, That the Bible, not as a sectarian book, but the revelation of God to man, be honored with a place in our common schools. Any system of education is radically defective which does not teach the moral law, and acquaint the pupil with the world's Redeemer.

8. To oppose infidelity of every form, and vindicate the Bible as the word of God. Let adulterers, and

robbers, and murderers believe and feel that no moral responsibility attaches to their conduct, that they can satiate their corrupt passions and execute their malicious purposes, and that if they can only manage to escape the action of criminal law, which they always think they can do, they have nothing else to dread, and you thereby "break the bands of God asunder," and sweep away the best safeguards of society. The following extract from the Bulletin of January 26, will exhibit Mr. King's position on this subject: "A notice of a dinner to be given in this city in celebration of a birthday of a noted infidel was sent to us yesterday for publication, which we declined. In our younger days, and before the down had fairly left our chin, we were foolish enough to entertain some doubts as to the truth of our holy religion. For some three years we devoted as much of our time, after leaving the counting-house, as we could, to the study of the Bible, and such books as treated of it. As before stated, we are not a member of any Church, but the result of that three years' study has been worth more to us than all the rest of our life, and we would not exchange the belief we have in the existence of God for all the wealth this world can produce." Will the masses of California, who have taken position with Mr. King on this platform, stand to it, and, by the utmost efforts of all legitimate influence, promote its ends? I may add, that the people, by their demonstrations against crime within the past fortnight, bear unmistakable testimony to the fact of the deep depravity of the human heart; and, secondly, to the necessity of retributive justice, as indispensable in promoting good government. If necessary in the administration of human governments, is it not necessary in the administration of the Divine? And if God certainly will "take vengeance on them that know not God, and that obey not the Gospel of our Lord Jesus Christ," then we see the necessity of reconciliation with God through the redemption which is in Christ Jesus, and a *reformation* of *heart* by the washing of regeneration and renewing of the Holy Ghost; by which these judgments are averted, and we become "fellow-citizens with the saints, and of the household of God." This is the only effectual means of reforming society. Violent outward remedies are sometimes necessary, as it is necessary to cut off a growing cancer; but the root is still there, and it will grow out again. The individual hearts composing society, must be regenerated by the grace of God, and thus the purified fountain will send forth a pure stream, "the good tree will bring forth good fruit." Will you admit these conclusions, and maintain these positions? We now hasten to our concluding task.

On the twenty-fourth day of February, 1843, a young man of twenty-one years, in Georgetown, D. C., sought in his closet, and there obtained, the pardoning mercy of God. Two days after his conversion he presented himself to the minister of the Methodist Protestant Church in that city, as a candidate for membership. His aged, pious parents had been members of that Church from its organization, and of the Methodist Episcopal Church for many years before. The minister said to our young friend that, as he was only known to the public as a haughty, worldly young man, he wanted him to give proof of the genuineness of his repentance by kneeling at the altar as a public seeker of religion. But the young man objected, saying that his sins were a matter between him and his God, and he had repented and obtained pardon, as he believed, and could not now go forward as a seeker of that which he had already obtained.

"Will you then," said the minister, "stand in presence of the congregation, and answer such questions as I may ask?"

"Yes," he replied; "I have no objections to that," and did accordingly, and was admitted.

His father had often said of him, because of his uniqueness of character, that "he was a drove by himself;" and it was a matter of doubt in many minds whether or not he would ever become a very docile sheep in the ecclesiastical flock, for he appeared to be a peculiar original thinker, and always spoke and acted as he thought. He went to the class to which his minister assigned him, and when it came to his turn to speak, he arose and said, in substance:

"Brethren and sisters, I have been listening attentively to all that has been said during the progress of this meeting, and I cannot say, upon the whole, that I am pleased with it. Brother —— there says that he is 'a great sinner, one of the vilest of the vile.' Now, I don't believe that, nor do I believe that he thinks so himself; and if a man were to come in here from the street, and say of him what he has asserted, he would hit him. And Sister —— there says that she 'sins dayly in thought, word, and deed.' I think, friends, when we come to class we ought to tell our experience, and tell nothing but the truth, and be consistent. I thank the Lord that I am not a vile sinner; but having sought and obtained the remission of all my sins, I stand here a free man in Jesus Christ, and, by the help of God, I don't intend to sin any more."

His very sensible, though severe remarks gave offense; complaint was made to the minister against the intractable young member, and he was called to answer. He was admonished and borne with, but a reciprocal dissatisfaction between himself and the Church was kept up, till finally he left it, but continued a God-fearing and praying man, and became a regular attendant, though not a member of the Presbyterian Church.

In the winter of 1843–4, he commenced the study of theology. In addition to his duties to his young family and his business as a banker's clerk, he applied himself diligently to the study of the Bible, Greek Testament, and other theological works for several years. That young man was James King, of Wm.

It is doubtless to this period of his history that he refers when he says, in the quotation we have given from an editorial in the Bulletin: "For some three years we devoted as much of our time, after leaving the counting-house, as we could, to the study of the Bible and such books as treated of it. The result of that three years' study has been worth more to us than all the rest of our life." There is also an evident allusion to the same period of his history, embracing his conversion, in his reply to "a friend," in the Bulletin of January 26, wherein he says: "Some twelve or fourteen years ago, *when we were moved to inquire into those things which make for our eternal good*, we read some Quaker books," etc.

Owing, probably, to his intense application to business and study, his health gave way, and he was led to seek its restoration on these healthful shores as early as 1848. These details, though not all matter of personal knowledge, were derived by me from a credible source, with permission to use them; and we have introduced them for the unpublished information they contain, and because they are characteristic of the man whose loss we mourn to-day.

In California he never joined any Church, assigning as a reason, that his peculiarities were such, that he could not, in every particular, affiliate with any one Church, while his general hearty good-will for the cause of God led him to help and encourage all. His catholicity on this subject is manifested in "a Church article" of the Bulletin, of February 25. We insert a short extract as follows: "We do not meddle with the creed of any Church. We view them all as working by different means and ways to the same end. At one end of the line we place the Roman Catholic, and at the opposite extremity the Unitarian Church, all the others being between these two." Then speaking of the progressive spirit of the Gospel, and the enlightening effect on the human mind, even of sectarian controversy, and the assaults of infidelity, he concludes in these words: "And the benefit of this advance in Christian knowledge is not enjoyed by the Church membership alone, but by those also in Christendom everywhere, who do not belong to any Church. By the preaching of the Gospel the world is better off to-day than it was some hundred years back. Morality is better defined, and gains strength just in proportion as the Church flourishes.

Let brotherly feeling between the members of the different Churches continue to be cultivated, as it is now getting to be, and Christians will bear and forbear with each other for the sake of the common cause, which may defy the assaults of the scoffer and the infidel."

Mr. King's moral character and conduct in California, though not decidedly Christian, would, I believe, have reflected more credit on any Church than that of a large proportion of her members, and his influence as a journalist has extended itself more widely, in that he was not tied to any party, political or religious. His case is unique, and does not furnish a precedent to be followed by the masses in regard to the duty of Church-fellowship.

But the question arises, Why should such a man come to such an end? Or, in the language of his bereaved widow, who manifested the most perfect self-possession from the time he received the fatal shot till his spirit departed, when, bending over his cold clay, she said, as she gave vent to the pent-up grief of her agonized heart, "I shall not disturb you now, my husband," and then exclaimed, "*Why did they kill you? Why did they kill my noble husband?*"

Upon that question we remark, that the essential cause of death exists in our moral relation to God. When we have finished the work assigned us in our probation, or filled up the measure of iniquity, the

Divine warrant is issued, "Set thy house in order, for thou shalt die and not live;" then any one of the ten thousand ostensible causes, or occasions of death, are sufficient to "loose the silver cord, and break the golden bowl." Hence God frequently suffers such to fall by the malice of their fellows. The murderous purpose, it may be, had long rankled in the heart, but while God's protecting shield was around the person of the victim, it could not effect its deadly aim. God has suffered the best of men in all ages, men of whom the world was not worthy, prophets, apostles, and even Jesus himself, to die by the hand of violence. And now the question recurs, "Why is this?" First to teach us, that the rewards of virtue and vice are not meted out in this world. Second, to give us an exhibit of the deep depravity of the human soul, that we may the better "justify the ways of God to men;" that we may realize the necessity of the provision of mercy in the Gospel; that we may appreciate and accept the atonement through the blood of Jesus as the only source of purification for our polluted hearts.

There are doubtless many other reasons justified by infinite wisdom. Why could not the prayers of the good avail to save the life of James King? There were certainly many prayers offered up, and much hope entertained. Christians have not lived right in California, and hence their faith is feeble, even when they have a positive promise on which to exercise it;

but in this case they had no promise. I said to Mr. King's boy of twelve years: "You must ask your Father in heaven, for the sake of Jesus Christ, to make your father well." He looked up at me instantly, saying: "I did." Soon after I made the same request of a younger brother, and in the same prompt manner he made the same reply: "I did, and I think he will do it." Thought I, here are "two agreed," and here alone is a lever of faith sufficient to raise a dying father, if they but had a fulcrum of Divine promise. I may ask with unwavering confidence for pardon, holiness, and heaven, for I have immutable promises on which to rest my faith. I may legitimately ask for worldly good, health, or the restoration of my friend, but unless the Holy Spirit reveal a Divine assurance in my heart, I have no certain foundation for my faith. I think it probable the good people of California had leverage power sufficient to have raised their hero, but they had no fulcrum. His work was done. I reported myself to James King immediately after he was shot. He asked me to stay by him, and taking his ring off his finger, he handed it to me, saying: "Take care of this." "For your wife?" "Yes." I studied his wants, and watched by his bedside day and night, with alternate intervals of rest, till he died. I will not go into a detail of the incidents of that most anxious week. He suffered much, but was invariably calm, patient, and self-possessed, till the morning of his departure, when he was a little flighty. He was a sincere penitent before God, said his "only peace was in the mercy of God through the merit of Jesus Christ." He asked me to pray with him at different times. On one occasion, rising from my knees, he said: "Give me your hand," and squeezed and shook it warmly. Speaking to him of the prayers of his boys, I remarked: "It is a great blessing for a man, in the day of peril, to have boys to pray for him." "Yes," said he, manifesting much pleasure in the thought. The only time I heard him use the name of his assassin, he said: "If I die, I don't want them to kill Casey." Manifesting the spirit of the command, "Love your enemies, and pray for them which despitefully use and persecute you." He several times spoke of "Charlotte and the little ones," with great affection. Many thousands in California will long mourn the loss of James King, of Wm., but we "sorrow not as those who have no hope."

CHAPTER XLVI.

THE FOUNTAIN OPENED.

On Sunday afternoon, the first day in June, 1856, as I was crying to a multitude on Washington-street, "Be ye reconciled to God," the Vigilance Committee arrested C. P. D., alias "Dutch Charley." From the place of his arrest to "Fort Vigilance," a full quarter of a mile, there seemed to be a flowing sea of excited humanity. When this great multitude began to disperse, I took a favorable position to catch the ebbing tide, and sung:

> "There is a fountain fill'd with blood
> Drawn from Immanuel's veins,
> And sinners, plunged beneath that flood,
> Lose all their guilty stains," etc.

And then spoke, in substance, as follows: "I have something of great importance to say to all Vigilance Committee men, and to all anti-Vigilance Committee men. The Prophet Zechariah, contemplating the saving power of the Gospel of Jesus, said: 'In that day there shall be a fountain opened in the house of

David, and to the inhabitants of Jerusalem, for sin and for uncleanness.' A purifying fountain for the polluted souls of sinners, blessed be God! Do we not need an application of some regenerating power, some cleansing process to make us what we ought to be—good citizens and good Christians? You all cry out, 'Reform! reform!' But do not imagine that by ridding the city of a few murderers, thieves, and 'ballot-box stuffers,' you will reform society. It may be necessary to cut off the excrescences—the great cancer warts of society, but unless an efficient remedy be applied to the roots, they will grow out again. Our business now is not with the rogues unhung which infest the city, but with you individually. To reform society, we must have the individual members composing society reformed. Have you not all sinned, and are you not unclean in the sight of God to-day?

"Are you an infidel? If so, do you not acknowledge some standard of right, some rule of moral conduct by which we should be governed? Have you lived up to that rule? Have you not violated it a thousand times? By the law of your own conscience you are a sinner before God, and have need of his pardoning mercy.

"Are you a Mohammedan? You have broken the laws of the Koran; you are verily guilty in the sight of God.

"Are you a Jew? What have you done with your ceremonial law and your sacrifices? Have you kept that true standard of morals which God gave to Moses, the Decalogue? O ye money-loving, Sabbath-breaking, God-forgetting, Christ-rejecting Jews, 'How shall ye escape the damnation of hell?'

"Are you a Catholic? When have you been to confession? You have made yourself so vile by willful, persevering rebellion against God, that the blessed Virgin would be ashamed to own you, if she were here to-day. You are so polluted that she would not look at you.

"Are you a Protestant? You believe the Bible, the Old and New Testament. You take the ten commandments as the great moral 'straight-edge' by which to find the irregularities of heart and life. Have you lived up to the line? Do you now love the Lord your God with all your heart, soul, mind, and strength? and do you love your neighbor as yourself? and do you, according to the law of Christ, 'love your enemies, and pray for them that despitefully use you and persecute you?' You know very well you do not. You love worldly pleasure, and sin, and yourself. You love those neighbors only who flatter your vanity and contribute to gratify your fleshly lusts. You don't love your enemies a bit; and as for praying for them, you do not even pray for your own poor soul. You are, indeed, guilty and defiled in the sight of God, and angels, and men. But if you do not all acknowledge the authority of the moral law, or if you do not perceive its spiritual import and application, I will try you by another test, another stand-point, from which you may try to measure the distance to that 'far country' into which you have wandered.

"Look here! Do you see this beautiful little girl?" at the same time patting the cheek and smoothing the hair of a lovely little girl of about three years, on the knee of a gentleman by my side. "On these little cheeks the blush of guilty shame never rose; the little heart that throbs within this breast beats in harmony with God; no stain of willful sin defiles her conscience. 'Of such is the kingdom of heaven.' What blessed innocence, humility, and confiding simplicity. Look at her." A streetful of sinners looked, and many wept. "Here is where you all once were; every one of you had then the innocence of this little girl, and were members of Christ's spiritual family. Had you died then, as some of your little brothers and sisters did, with them you would have gone to dwell with Jesus and holy angels; but where are you now? O how far you have gone from home. I might justly challenge the mathematical skill of the angel Gabriel to compute the distance of your flight, or number the steps of sinful departure from infantile innocence to your present wretched condition. O, how your sins have multiplied, especially since you came to California. You have been all this time engaged in the miserable business of 'treasuring up unto yourselves wrath against the day of wrath and revelation of the righteous judgment of God.' Now what are you going to do about it? Something must be done, and done immediately, or you are ruined forever. You have to get back to the state of this little girl. 'Except ye be converted and become as little children, ye shall not enter into the kingdom of heaven.'"

I then pressed the invitation to all to come to the "fountain opened," the fountain of redeeming mercy in Christ, by a variety of arguments and illustrations, as the Spirit gave me utterance. Good order and great seriousness prevailed throughout the assembly. Eternity will reveal the fruit.

CHAPTER XLVII.

THE MISSIONARY TO NINEVEH.

On the eighth of June, 1856, at half past four in the afternoon, I took my stand on the steps of the Pacific Mail Steamship Company's Office, in full view of "Fort Vigilance," the great center of attraction in this city for some weeks past.

When a boy, I sometimes went hunting, and always tried to go where the game was. My commission reads, "Go ye into all the world, and preach the Gospel to every creature;" I therefore always look out to see where the greatest number of those creatures congregate, and try to take every advantage of "wind and tide," so as to bear the message of mercy to the greatest possible number, under the most favorable circumstances the case will allow.

On this occasion I got the "windward" of a large, attentive audience. The subject of discourse was the great reformation in ancient Nineveh, under the preaching of Jonah; just such a reformation as we need here in San Francisco.

Speaking of the instrumentality by which the great work was brought about, the discourse ran as follows:

"God wanted a missionary to go and preach to the Ninevites, and called to this responsible work a Hebrew by the name of Jonah, saying to him: 'Arise, and go to Nineveh, that great city, and cry against it; for their wickedness is come up before me.' But Jonah, like Moses, his great ancestor, begged to be excused. He probably, in effect, said: 'Lord, I am "slow of speech," and cannot succeed as a preacher; the distance is great, and I know not that I should live to see Nineveh; and then, if thy servant should stand up in the streets of that wicked city, the people would stone him to death. And even if I should get there safely, and preach successfully, and the people should repent, thou wilt not bring upon them the judgments of thy word. "For thou art a gracious God, and merciful, slow to anger, and of great kindness, and repentest thee of the evil." So the people will call me a lying impostor, and perhaps take away my life; and then relapse into idolatry worse than ever. Lord, I cannot go to Nineveh. Be merciful to thy servant, I cannot go.' When he found that the Lord would not release him, he determined to get out of the way. Some young men, when called to preach the Gospel, suddenly take a notion to get married, and go into some complicated co-partnership business; so binding themselves that they cannot honorably withdraw for a term of years. But 'Jonah rose up to flee into Tarshish from the presence of the Lord.' Tarshish is believed to be the same as Tartessus, in Spain, near the Straits of Gibraltar, at the opposite point of the compass from Nineveh. Darkness immediately follows disobedience. Jonah thought: 'O! if I can only get a passage to Tarshish, the Lord will not find me, and I will be free from this dreadful responsibility.' So many of you thought when you started for California. You said in your hearts, 'I have been teased so much about going to Nineveh, or doing some other unpleasant business for the Church, I'll go off to California, where I can do as I please, without let or hinderance. I'll have done with father's rebukes, and mother's entreaties and tears, and this everlasting preaching about hell and judgment. I'll leave all responsibility behind, and I'll be a free man for once in my life.' And lo! ye find that God is in California too, and he has been speaking to you by various providences, in thunder tones; and the same unwelcome Gospel truths and threatenings of future retribution peal forth unexpectedly upon you, even from the corners of the streets; and you are so disappointed in the felicitous freedom of California life, that you begin to feel, 'O that I had a mother's prayers and counsels to guide and comfort me as in other years.' I will tell you, my friend, God has not done with you yet. You had better look well to your ways.

"So Jonah went down to Joppa, the nearest sea-port to Jerusalem, and, walking along the beach, he saw a sign on a ship: 'For Tarshish; will positively sail at the sixth hour to-day.' Good, thought he, I've just come at the nick of time. Though guilty and downcast, he now put on rather a hopeful, courageous appearance; and though a little frightened at the sound of his own voice, he hailed pretty well for a 'runaway,' 'Ship a-hoy! Is the captain aboard?' 'Yes,' replied the officer of the deck, 'walk aboard, sir. You'll find the captain there in his office, sir.' With a slow, hesitating step, Jonah entered the captain's room. 'Well, stranger,' said the captain, 'what can I do for you?' 'I—I—I wa-want to go to Tarshish, sir.' 'Very well, sir; we sail to-day at the sixth hour.' 'What is the fare, captain?' 'Thirty pieces of silver.' So he paid the fare, and went down into it, (the steerage,) to go with them unto Tarshish, from the presence of the Lord. Jonah immediately 'turned in,' and fell into a sound sleep. When conscience is gagged by violence, and its voice drowned by the whirlwind of passion, you generally get one sound, though guilty, sleep, before it is able to rally and reassert its injured prerogatives. But rally it will; and, like a lioness robbed of her whelps, will pounce upon you, and tear you asunder. I had rather fall into the clutches of a grizzly bear, than into the foldings of a guilty conscience, fully awake. They had been under sail but a few hours, when 'the Lord sent out a great wind into the sea, and there was a mighty tempest, so that the ship was like to be broken. Then the mariners were afraid, and cried every man unto his god.' These poor heathen sailors had not learned the modern art of explaining God out of the universe, by some metaphysical disquisition on the laws of nature. These were the facts in their simple minds: The storm rages, the gods are angry; there is a cause. We have sinned; retributive justice is awake. Some one of us has committed a very great sin. Who is the guilty man? Let him be delivered over to justice, that we all perish not. 'Captain, where is that suspicious-looking stranger that engaged passage yesterday?' 'I don't know. I'll see.' The captain ran 'down into the sides of the ship,' to Jonah's 'bunk,' and pulled off the blanket; and there 'he lay, and was fast asleep.' And the 'shipmaster' cried, 'What meanest thou, O sleeper? Arise, call upon thy God, if so be that God will think upon us, that we perish not.' By the time Jonah got on deck, the crew had prepared to cast lots, and thus detect the guilty man. 'And they said every one to his fellow, Come, let us cast lots, that we may know for whose cause this evil is upon us. So they cast lots; and the lot fell upon Jonah.'

"Poor deserter. You may imagine how dreadfully sea-sick he was, and how his conscience lashed him, and how his guilt exposed him to the scrutiny of the crew, and to the vengeance of God and man. So they pressed the question, 'Stranger, tell us what thou hast done to provoke the gods, and bring all this evil upon us? What is thine occupation? and whence comest thou? What is thy country? and of what people art thou?' And Jonah told them all about it, his only relief being in confessing the truth.

"A young man, whose name I need not mention, left the Eastern States and came to California for the sole purpose of concealing his sins and shame. After wandering a few months in the mountains, he came to San Francisco sick. In a dreary room in Dr. Shuler's hospital, I saw him struggle with death. The flickering light of a small lamp seemed only to make the darkness more perceptible, and render the scene more gloomy. His piteous groans echoed in the depths of my soul. I urged him to pray, but, said he, 'It's no use. It is too late now; I am lost; but I want to tell you what a sinner I have been. I have concealed my sins till they have become a consuming fire in my heart. I feel some relief in telling what I have for years been trying to conceal.' He then detailed his crimes, and told how he had deceived his parents, resisted the entreaties of his pious aunt, and grieved away the Spirit of God.

"Sinner, those deeds of darkness you conceal with so much care, and would not have me spread out before this audience for a fortune, and would not have your mother know for all the gold of these mountains, will so sting your conscience in death that you will most gladly seize the small relief of open confession. 'Then said they unto Jonah, What shall we do unto thee, that the sea may be calm unto us? For the sea wrought, and was very tempestuous. And he said unto them, Take me up, and cast me forth into the sea; so shall the sea be calm unto you; for I know that for my sake this great tempest is upon you.'

"But those generous-hearted tars, though they had lost their freight on his account, and were exposed to such peril, were unwilling to cast the guilty man into the sea, but 'rowed hard to bring the ship to the land,' but the tempest raged with increasing violence. Then they all called on Jonah's God, saying, 'We beseech thee, O Lord, we beseech thee, let us not perish for this man's life, and lay not upon us innocent blood.' When they saw there was no alternative but to drown the preacher, or all perish together, the captain said, 'Mr. Jonah, we are sorry for you, but you see our fix.' 'So they took up Jonah, and cast him forth into the sea.' The sailors watched him with tearful eyes as he struggled amid the raging billows, till the cry simultaneously arose from the whole ship's company, 'He's gone! he's gone! There is an end of that lubber.' But God, who was thus educating Jonah for the great work of his mission, 'had prepared a great fish to swallow him up.' It don't matter whether this fish was a whale or a shark; as the whole affair was a miraculous interposition of Providence, God could easily fit either for his purpose, or make a new fish for the occasion; at any rate, Jonah was swallowed, and the sudden transition seemed to realize to him all his forebodings: 'Dead, and in hell at last, just as I expected, and as I deserved. In the very belly of hell.' A little reflection, however, and a close examination, convinced him that he was still in the body, and he was astonished to find that he was not dead. But he never changed the name of his lodgings. It stands in the book, 'the belly of hell,' to this day. He could imagine nothing worse than that. 'Then Jonah prayed unto the Lord his God, out of the fish's belly.' An unpropitious place for repentance, one would think. 'From the bottoms' of the submarine mountains of the Mediterranean Sea, at a depth of a thousand fathoms, he prayed, 'O Lord, deliver me from this hell, and I'll go to Nineveh. O Lord, be pleased to deliver thy servant, and I'll go. O Lord, I'll go anywhere, and do anything thou commandest. I will serve thee.'

"Is it not probable that Nehemiah, or whoever it

was that composed the one hundred and thirtieth Psalm, copied Jonah's prayer? 'Out of the depths have I cried unto thee, O Lord. Lord, hear my voice: let thine ear be attentive to the voice of my supplications. If thou, Lord, shouldest mark iniquities, O Lord, who shall stand? But there is forgiveness with thee, that thou mayest be feared. I wait for the Lord, my soul doth wait, and in his word do I hope. My soul waiteth for the Lord more than they that watch for the morning: I say, more than they that watch for the morning.' God hearkened to the prayer of Jonah, and, on the third day, commanded the great fish to set his passenger ashore. The gladness of Jonah's heart when he jumped ashore, who can tell? His experimental song of praise to God, which he sung in the ecstasy of his redemption, is recorded in the second chapter of the book bearing his name.

"'And the word of the Lord came unto Jonah the second time, saying, Arise, go unto Nineveh, that great city, and preach unto it the preaching that I bid thee. So Jonah arose, and went unto Nineveh, according to the word of the Lord.' By what route or conveyance we are not informed; but he did not deviate from his 'orders.' He dared not, even if he had been inclined, lest a whale or some other monster should seize him.

"Let every man whom God has called to preach

the Gospel feel, 'Woe is me, if I preach not the Gospel.' Jonah had his diploma, and I do not wonder at his success as a preacher. O that God, in his mercy, would, by some process, educate a few Jonahs, and send them to wake up the careless sinners of this wicked country."

CHAPTER XLVIII.

THE DOWNFALL OF THE HAMAN FAMILY.

On Sunday afternoon, the twenty-second of June, 1856, at the corner of Sacramento and Liedsdorf streets, I announced as my text, to a very large audience, "The adversary and enemy is this wicked Haman."

The day before, Saturday, the twenty-first of June, was a day of great excitement in the city. Judge T., one of the supreme judges of the State of California, stabbed Sterling A. H. The great bell of the Vigilance Committee struck three times, and in a moment the whole city was in commotion. All business was suspended, stores were closed, dray-horses were stripped of their gear, leaving the loaded drays in the streets, to join the cavalcade. In half an hour nearly the whole force of the Vigilance Committee, numbering six thousand men, were under arms. Long columns of muskets, bayonets, and sabers gleamed in the sunlight; but all in solemn silence. No drum, no shouting, naught but the stern command of the officers. The only distinguishing badge of this army was a small piece of white ribbon or cloth, tied in a button-hole of their coats, or vests, if they had no coat on. One fellow, as he ran to get his musket, calling to mind the fact that he did not have his badge, turned a corner, tore a strip off his shirt, tied it into the lapel of his coat, and on he went.

Judge T. had taken refuge in the "Armory of the California Blues," the head-quarters of what was called "the law and order party." The armory was immediately surrounded by detachments of the vigilant army, who demanded the prisoners and all the fire-arms and munitions of war contained in the building. The doors were opened by the surrendering party, and the "Vigilants" took possession. On the bulletin board inside were seen posted notices for a grand parade of the law and order forces, to be on Sunday, the twenty-second, at ten o'clock A. M., and a review of the army by Gen. V. E. H. Judge T. and some other prisoners were placed in two close carriages, the grand cortège formed around them, and marched in solemn procession to "Fort Vigilance," on Sacramento-street. The front ranks consisted of a large body of infantry, next in order the carriages containing the prisoners, next several dray loads of muskets and cartridge-boxes, the trophies of war, followed by a large guard of infantry. The cavalry brought up the rear. After conveying the prisoners to the "fort," detachments were ordered out to take possession of all the armories and arms of the opposing party. There were three more besides the one they had just taken. The whole was accomplished, and about ninety prisoners marched in irons to prison, without collision or bloodshed. Most of the prisoners were discharged the next morning from custody. In a few hours the surface of society was calm, business was resumed, and gentlemen, ladies, and children were seen promenading the streets in all directions, as though nothing had occurred.

A mass-meeting of about ten thousand citizens, held a few days before, endorsed the position and operations of the Vigilance Committee; and it is confidently asserted by a majority of the public journals of the city, that nine tenths of the inhabitants of the city and of the state approve the action of the committee, in view of the wrongs this community has so long suffered, and feel great security of life and property under their administration. I always, so far as I know the right, declare my approval of the right, and condemnation of the wrong; but I belong to no party, and take no active part on any exciting party question, extraneous to the one appropriate cry of my calling, "Behold the Lamb of God, that taketh away the sin of the world."

The foregoing is a hasty review of the surroundings of the preaching occasion to which I have invited attention. The story of Esther is familiar to all Bible readers. I will, therefore, simply note a few points in the application of the discourse in question: I made Mordecai "the personification of that stern religious principle which constitutes the integrity and stability of the Church in all ages. He worshiped God, and God only; he recognized the authority of the 'higher law,' and never hesitated between the alternatives of 'obeying God or man.' And yet he sat at the gate, comparatively unknown, poor, and despised. Esther was our representative of active virtue, implying spiritual understanding, submission to the will of God, unwavering faith in Jesus Christ, and all the manifest graces and fruits consequent upon the exercise of it. She is very nearly related to Mordecai. Bigthan and Teresh were representatives of a large class of murderers, gamblers, and 'ballot-box stuffers.' They aspired to be princes in the city of Shushan. They have constituted the aristocracy of the city of San Francisco, moving in courtly pomp and splendor. Everybody knew them to be nonproducing, worthless men in society; but it was not suspected that they would put on the livery of the law, subvert the reign of justice, clandestinely trample under foot the elective franchise, and other sacred rights of American freemen. Bigthan & Co. despised Mordecai, would take no notice of him, and supposed he took no notice of them; but Mordecai is always a loyal subject, and a true friend of good government, and watches with ceaseless vigilance the insidious movements of the Bigthan fraternity. He thus detected their secret plots, and through the influence of Esther, his kinswoman, brought them to justice.

"The avowed object of the Vigilance Committee is to clear this city of the whole clan of Bigthans and Tereshes. Mordecai has been marking their movements for years, and has testified against them. Esther has a voice in the counsels of the Committee. Like an angel of mercy, she hovers over the executive in their deliberations. They have received wise counsel from her lips. But should they succeed in exterminating or banishing all the Bigthans of the land, still Haman remains. We have to look out for him, for he has great wealth and influence; and though he will not now oppose the counsels of Esther, he is a most dangerous man. Haman is an infidel; he repudiates the word and authority of God. He is a tyrant, he has no regard for the claims of suffering humanity. He is an enemy of all righteousness, because not consonant with his lascivious passions and plans. He is a political demagogue, who would sacrifice a whole nation of Mordecais on the altar of his ambitious pride, and would pay 'one million one hundred and nineteen thousand pounds sterling' for the accomplishment of his ambitious and malicious purposes.

"I heard a man, yesterday, say that he had expended ten thousand dollars to be elected sheriff, and was disappointed after all. Haman is the fellow, sitting in the counsels of the Vigilance Committee, side by side with Esther the queen, that will give us trouble yet. He is a most wily politician. Mordecai will have to sit at the gate in California for several years to come, before we shall be able to dispose of this dangerous foe. He seems very kind and pliable now; but, as he acquires influence, he will the more despise Mordecai, 'and plot against the just, and gnash upon him with his teeth.' But let Mordecai maintain his fidelity to God, and do his duty in California; let Esther maintain her purity of heart, and her activity in Christian enterprise; and let all the people of Mordecai and Esther fast and pray, and God will make the counsel of Haman like that of Ahithophel. He will lift up the head of his servant Mordecai. Don't be discouraged, my good fellow; 'commit thy way unto the Lord, trust also in him, and he will bring it to pass. He shall bring forth thy righteousness as the light, and thy judgment as the noon day.' And God shall 'bring it to pass' so unexpectedly, and so opportunely, that you will exclaim with David the king, 'When the wicked, even mine enemies, and my foes came upon me to eat up my flesh, they stumbled and fell.' Just as they were about to devour me 'they stumbled and fell, and I escaped.' See the displays of God's wise providence in 'Shushan the palace.'

"Haman and his party exulted in his promotion, as the sole guest, with King Ahasuerus, at the queen's banquet, and regarded that as an unmistakable indication of the final success of all his ambitious schemes. But there's that stubborn Mordecai at the gate; he cannot longer be tolerated. Mrs. Haman, true to the class of Jezebels and Herodiases to which she belongs, the very antipodes of Esther, suggested the happy expedient. 'Let a gallows be built seventy-five feet high,' and go early to-morrow morning, and obtain from the king a death-warrant for Mordecai, and hang him, (or impale him, rather,) and then thou canst enjoy the banquet of the queen. Strange as it may seem, the king could not sleep that night, and said to his scribe, 'Bring hither the book of records of the chronicles, and read before me.' The chronicles of Eastern kings were written by the best poets, in measured verse, so that the reading of them was very entertaining; much more so, we should think, in view of their historic worth, than the novels of modern days. Providentially, the scribe read 'where it was written that Mordecai had told of Bigthan and Teresh,' and was thus the means of saving the king's life, and the king said, 'Stop, sir; what honor and dignity hath been done to Mordecai for this?' 'Then said the king's servants that ministered unto him, There is nothing done for him.' And the king said, 'Sentinel, who is in the court?' 'Behold, Haman standeth in the court,' was the reply. 'Tell him to come in,' said the king. So Haman came in, and the king said unto him, 'What shall be done unto the man whom the king delighteth to honor?' 'Now Haman thought in his heart, To whom would the king delight to do honor more than to myself? I alone was his guest yesterday.' 'And Haman answered the king, For the man whom the king delighteth to honor, let the royal apparel be brought which the king useth to wear, and the horse that the king rideth upon, and the crown royal which is set upon his head. And let this apparel and horse be delivered to the hand of one of the king's most noble princes, that they may array the man withal whom the king delighteth to honor, and bring him on horseback through the street of the city, and proclaim before him, Thus shall it be done unto the man whom the king delighteth to honor.' Then the king said to Haman, 'Make haste, and take the apparel and the horse, as thou hast said, and do so'—(Yes, thought Haman, put it on myself, of course I am the man. This head of mine shall bear the crown royal at last)—'and do so to Mordecai the Jew, that sitteth at the king's gate.' 'My lord, O king! live forever, thy servant—' 'Not a word, sir; go,' said the king, 'and let nothing fail of all thou hast spoken.'

"Did you ever in all your lives see a man so crest-fallen? Judge T. did not feel worse yesterday when arrested by the Vigilance Committee. So here comes Haman, with the royal apparel and the crown, leading the king's horse to the gate. There sits Mordecai," pointing to Captain E., who has proved himself a worthy representative of Mordecai for six years in California, "stern in his integrity, but how greatly astonished, when his old enemy said, 'Mordecai, stand up, sir, and allow me to put upon you these royal robes and this crown. Mount the king's horse, sir.' And down the street they went, Haman leading the king's charger, and with choked and broken utterances, proclaiming, 'Thus shall it be done unto the man whom the king delighteth to honor.' The fate of the Haman family is sealed. The redemption of Mordecai and his people secured.

"'God moves in a mysterious way
His wonders to perform.'

"Only let Mordecai and Esther do their duty in California: let the infant Church of Jesus in this wicked land, 'stand in the ways, and see and ask for the old paths, where is the good way, and walk therein,' 'obeying God rather than man,' though now sitting at the gate in rags, and the time will come when Mordecai's God will say to her, 'Arise, shine, thy light is come, and the glory of the Lord is risen upon thee.' She shall then come up out of the wilderness, 'fair as the moon, clear as the sun, and terrible as an army with banners.' And from all these streets, and our beautiful valleys, and from hill-top to hill-top, nay, from the coast-range to the snow-capped summits of the Sierra Nevada mountains, one universal California shout shall arise, 'Halleluiah! The Lord God omnipotent reigneth.'"

CHAPTER XLIX.

LETTERS FROM HOME.

There are many objects and places of attraction in California. Indeed, it is altogether a very attractive country, as its population of three hundred thousand, attracted from all parts of the world in the space of eight years, will clearly prove. There is a charm in its climate, its scenery, bays, rivers, valleys, mountains, and ocean; its varieties of production, mineral and vegetable, and its game, fowl, fish, elk, deer, grizzly bears, etc. The great magnet is its rich deposits of virgin gold in banks that never fail, and on which every man may draw. Only make a run on them, and get them into liquidation, and they will pay all the better. But the greatest local attraction, of the heterogeneous masses here attracted, is the post-office. Thousands of men here, who never were absent from their wives and children a week at any one time, till they started for California; thousands of young men, who scarcely were ever out of sight of the smoke of their mothers' chimneys till they bade good-by to "the old folks at home," to try

POST-OFFICE IN 1850.

their fortunes in the land of gold; hundreds of young lovers, bound by sleepless affection and plighted faith to virgins beautiful and lovely, to whom they would certainly return in two years, which was all the time any decent man could ask to make a fortune in California. Six months would probably realize all their hopes, but to be certain of no disappointment to the fair ones, the time was set for two years. How desolate the hearts of these different classes of men, in the absence of all those objects of home attraction and affection, in this vast social Sahara. The only substitutes for them were the little drops and glimpses of social life and light obtained through the post-office. A view of the office at San Francisco, with which I have been familiar for more than seven years, will describe, in the main, all the post-offices of this coast. At first they had "two windows of delivery." One was for the "navy and army, the French, Spanish, Chinese, clergy, and the ladies." All the rest of mankind in California were waited on at the other window, provided they had time and patience to take their turn, and work their passage to it. Every man had to wait his turn, as the country mill boys used to do. The line of anxious faces, single file, was, on the arrival of every mail, from one to three hundred yards long. To travel from the rear end to the long-desired "window" was a work of from one to five hours. This long line hardly ever began to shorten for half a day after it was formed. Its slow travelers, never in such a hurry before, making from one to two steps in their journey every minute, were entertained and fed, or bored by the newsboys, fruit boys, pop corn boys, and candy boys. The boys, who have so hard work to keep up with our fast men, or get a hearing in the streets, seem always to feel that they have a rare advantage over the men of the line, and improve it to the best of their skill. The slow travelers are weary, hungry, have calls of pressing importance, and their time more valuable than gold, but they must not break rank, or they will lose their turn, and have to begin again. Men sometimes bought a chance near the window for five dollars, and got their letters without much delay, while the speculators in chances went back and commenced anew. To look at the anxious countenances of men at the windows was painfully interesting. One man gets a letter, and immediately breaks it open, expecting "news from home," but, lo! it is a letter of introduction from some man he never saw, who has "taken the liberty of referring a particular friend" to him for information, and the "particular friend not meeting with him so soon as he expected, dropped the letter into the post-office." He tears up his only letter, and hopes never to be introduced to that "particular friend." Another is waiting in great suspense, but the postmaster says: "Nothing for you, sir."

"Please, sir, look again," says the expectant.

"Nothing for you, sir."

Turning away, he says: "I came round Cape Horn, and they were to commence writing after I had been out a month, and now it is eight months, and I haven't got a letter."

The next one gets a letter, and breaking it open, as he turns away, you see him trembling till black with agonized emotion. You at once know that some dread bolt from that letter, but little less powerful than a thunder-bolt, has struck him. You see no tears, for they seem to be frozen up in their fountains. The only utterance you hear from his lips, broken and involuntary, as he retires from the crowd, is: "O, my God, she is dead!"

The next man awaits his portion with trembling. He gets a letter, pays forty cents postage on it, and breaks it to get the news from home. "Pshaw!" says he, "I think a fellow writing to know whether he had better come to California, might pay the postage on his letter. I shall write him to stay at home."

Another standing at the window says: "I have not received a letter for six months, and I expect it will be just so this time."

"Perhaps," said I, "you do not write to your friends?"

"Yes, I do," said he, "but I can get no answer."

"Nothing for you," says the post clerk to him, and he turns away with a sigh.

A man takes out a letter, and reads, and presses it to his lips, and reads on, and kisses it again and again. His tears break through a "windrow" of smiles on his face. It is from his dear wife; and John, and Mary, and Lizzie have all added a postscript.

"In the course of human events," the post-office was moved down to the "Portsmouth House," on the west side of the Plaza. There, with a great increase of room, the windows were multiplied. The navy and army had a place to themselves. The French, Spanish, and Chinese had their window, while the ladies and clergy still kept company to the same window. The great undistinguished masses were divided into classes by the letters of the alphabet.

All whose names commenced with a letter included between A and D fell into the A and D line. Another class for the window of E and H; and so on through the alphabet. This was quite an improvement on the old system. By and by we had "boxes," in which the letters could be seen from the outside. "Box rent" was quite an "item," but that was nothing to a man anxious to get "letters from home." Then, again, we had boxes with doors opening on the outside, and the renter of the box carried the key, so that he could open it whenever he pleased. Still a great many have to take turn in the line, and all the improvements in the office could not supply the disappointed with any equivalent for the expected "letters from home" they did not receive, nor extract the death-shocks from those which bore the "black seal." An anxious man, who had taken his turn in the line, expecting a letter from his wife, received a letter, but O misery! it was from a crazy woman, who had fallen in love with some man, and had written to this expectant friend to send her dear Mr. —— to her, or she "certainly would die and be lost forever." A poor lover, who has had "bad luck," and has not been able to return at the time appointed, is waiting at the "window" in great suspense. He hopes to get a letter from his dear S., telling him that she loves him still, and will wait till he can "make a raise," or have him without the "raise." He receives a letter from a friend, informing him, that, "Alas! alas! his S. is married!" Poor fellow, he feels that what little is left of himself is hardly worth saving, and hence throws himself away.

A true-hearted girl did write to her lover to the last, and when her lover's trunks were sold at public auction to pay his funeral expenses, these letters, too sacred for such an exposure, nevertheless bore testimony to her unwavering affection. "They were letters from home," and they soothed a dying man.

Again to the window. Another poor fellow, as he turns away in deep disappointment, says: "I have not received a letter from my family for two years. Thinking it might be the fault of the Mountain Express men, I have come down here, three hundred miles, and have spent one hundred and fifty dollars to try to get one letter from home, and I can't get it. I'll just quit writing! It's no use!"

The next man gets a letter; breaks it, reads and laughs. Reads and laughs again, seemingly unconscious that anybody sees him; except, indeed, he imagines himself really in the presence of those with whom his soul is evidently conversing.

Nathan Withers, a seaman, who had not received a letter from his family in Scotland for seven years, wrote them to address their letters to my care. In due time, the long silence was broken, the letter came; and the old tar, unused to weeping, wept for gladness, as he read, from the hand of his wife, about his children that had grown up in his absence. They had received his letters, and money for their support regularly, and had written him; but, in his frequent changes, he had not received a letter for seven years. The post-office has usually been closed on the Sabbath in San Francisco, from the first, except when the mails arrive on Saturday night, too late for distribution. On one of these occasions, the "general delivery" was opened at the hour I was by appointment to preach on the Plaza, in the immediate vicinity where the lines formed and passed. As I was about to commence the announcement of my "news from a far country," a man came up in a hurry, and said to me, "Is this the line to the A and D window?" "I don't know, sir," replied I. "I am about forming a line, sir, to travel to the kingdom of heaven. I shall be very glad to have you fall into our line, sir, and go with us." "I don't wish to go there yet, sir," said he; "I want my letters from home."

CHAPTER L.

PATRIOTIC PERSUASIVES TO BE A CHRISTIAN.

At the corner of Sacramento and Liedsdorf streets, on Sunday, the thirteenth of July, 1856, I announced as my text, "Almost thou persuadest me to be a Christian." The audience was large and orderly. I gave a very brief history of St. Paul's imprisonment and trials: "Arrested by a Jewish mob in the temple; rescued by Colonel Lysias, with a troop of Roman soldiers from the Castle of Antonia; arraigned before the Sanhedrim, where they would have given him 'law and order' to the death, but for the timely interposition of Colonel Lysias, who afterward sent him under an escort of four hundred and seventy soldiers, infantry and cavalry, to Cæsarea, the seat of Roman authority in Palestine, to be tried before Judge Felix. The prosecution was conducted by the Hon. Tertullus. St. Paul, who had been educated at the feet of Gamaliel, a doctor of laws, and afterward graduated in the school of Christ, was lawyer enough to plead his own defense; and he did it in a masterly manner, refuting most conclusively the threefold charge of sedition, profanation of the Temple, and heresy. And if his honor, Judge Felix, had been an honest judge, he would have discharged the prisoner at once, by an honorable acquittal; but for reasons, which even that long-cued Chinaman could detect, he kept him in prison two years, and then delivered him over to his successor in office, Judge Festus. First, he wanted to please the Jews, by whom his vanity was flattered; and, secondly, he wanted St. Paul to pay him money. A Chinaman, up country, when a fellow-Chinaman was arrested for a murder, was asked what he thought would be the fate of the prisoner. 'O!' said he, 'he get free; he no hang. He just same as one Melican (American) man. *He got money.*' If St. Paul had been the same as 'one Melican man,' he could have been liberated immediately; for the judge, like some Californian judges I have heard of, had his price, and was in the market waiting for a bid. But St. Paul, a poor, despised missionary of the cross of Jesus, had no money; for all the extra change he had received on his circuit the year preceding, he had just given away to the 'poor saints at Jerusa em.' But if his friends had furnished the money, Paul disdained to pay a premium on the cupidity and corruption of the judge, or accept of liberty on such conditions. From Judge Festus he received the same kind of treatment; and after he had taken an appeal to the Supreme Court at Rome, Festus said to his guest, King Agrippa, and the audience assembled to hear Paul: 'I have determined to send the prisoner to Rome; of whom I have no certain thing to write unto my lord. Wherefore I have brought him before you, and especially before thee, O King Agrippa! that after examination had, I might have somewhat to write. For it seemeth to me unreasonable to send a prisoner, and not withal to signify the crimes laid against him.' So it would seem, Judge Festus, to any man who had one grain of common sense, especially after the man had been in prison for more than two years; and I wonder you did not think of that when his case was tried before you.

"St. Paul, in chains, preached to the assembled audience, and especially to the illustrious quaternion, King Agrippa, and his vile sister Bernice, and her sister, Mrs. Festus, and the honorable judge himself. Paul's arguments were unanswerable, and such was the persuasive power of his eloquence that King Agrippa interrupted the preacher by crying out, 'Almost thou persuadest me to be a Christian.' Agrippa was not almost a Christian, but almost persuaded to be a Christian.

"In the further discussion of the subject, I will consider:

"I. What is a Christian?

"II. What are the persuasives to influence the will to accept of Christ, to become a Christian?

"To suit our present purpose, we will present the persuasives under two heads: 1. Patriotic Persuasives. 2. Persuasives purely Spiritual, such as were employed directly by St. Paul.

"We will here note a few points and illustrations, under the head of Patriotic Persuasives, as presented on Sacramento-street.

"You all profess to be patriots, do you not? Yea, most of you profess to be reformers. Your connection with the Vigilance Committee is for the avowed purpose of reformation. Whatever, therefore, will most directly effect the desired end, should be matter of great interest to you. The strength, prosperity, and permanence of a nation do not consist in her navies and armies, nor her walled cities and fortifications, nor her colleges, academies, and public schools. These are all necessary appliances of protection and development, and *evidences* of a nation's strength, but not the *basis*, nor *source*, nor *primary conditions*. 'Righteousness exalteth a nation, but sin is a reproach to any people.' What is the cause of all the evils so much complained of in this city? What is this rum-selling, and drunken debauchery, and gambling, and theft, and bloodshed, and corruption in office, and ballot-box stuffing, but the development of *sin* in the *hearts*, and its corresponding manifestation in the *lives* of those various characters? All our degradation and imbecility, individually and collectively, proceed from the same source. It is the opposite of all this, righteousness, experimental godliness alone, that will purify and exalt society. This is the great conservative bond that constitutes the strength and integrity of any nation. The connection between the cause and effect, maintained by this proposition, is not seen by superficial observers, but you have an illustration of it in the declaration of God in regard to Sodom: 'I will spare the city for ten's sake.' This is the salt that preserves society from utter moral putrefaction. The health of a people consists of the health of the *individual* members composing society. If, therefore, we sincerely desire to see a reformation in this city, and to see society elevated and established on a permanent moral basis, we must earnestly apply ourselves to the work of *personal* reformation

"Now, how far do the avowed purposes of the Vigilance Committee go to effect this? I believe that all they propose to do, besides the moral effect of an expression of the popular voice against certain sins, which, to be sure, is very important, is 'to clean the Augean stable.' I am not very familiar with that stable, but it doubtless needs cleaning, and we wish them good success in the dirty job they have undertaken, and that all the moral nuisances of the city will be cleared out. But how far will that go, however necessary, as a preparative toward purifying and elevating society? That is but removing the rubbish.' Nehemiah's men had to clear away 'the rubbish;' but had they stopped there, the walls never would have gone up. As the rubbish is cleared, we must lay the foundations of the temple of truth on these shores, deep and broad, and then go up with the walls, and, trusting in God, 'the mountain' that now obstructs 'will become a plain,' and the temple, built up of 'living stones,' shall be completed, and our Divine 'Zerubbabel shall bring forth the head-stone thereof with shoutings, crying, Grace, grace unto it.' But how are we to proceed in this work? We have a good model in the great reformation of ancient Nineveh, under the preaching of Jonah: 'The people of Nineveh *believed* God, and proclaimed a *fast*, and put on *sackcloth*, from the *greatest* of them even to the *least* of them.' What a worthy example the old king set for his people! The influence of high officers of state is immense, for good or for evil. 'The wicked walk on every side, when the vilest men are exalted.' How true to the experience of California! The king proclaimed a fast, not as some hypocritical governors we have heard of, who, after proclaiming a solemn fast for the people, spend the day in debauchery. 'The king arose from his throne, and he laid his robe from him, and covered him with sackcloth, and sat in ashes.' The people imitated his example, and they all 'cried mightily unto God, and *turned every one from his evil way.*' That is the kind of reformation we want here in San Francisco.

"Now, you who have families here, when you go home to-night say to your companion, 'My dear wife, God has intrusted us with these infant germs of immortality, and their weal or woe, for time and eternity, depends mainly upon the training we give them. We have not done our duty toward the souls of our children, our own souls, nor the souls of our neighbors. Now let us read a lesson in the Bible to-night, and try to pray with our children.' 'O, but,' says one, 'I don't feel like it, and it would be mockery to attempt it.' Make an honest effort upon the decisions of your judgment, with or without feeling; the feeling will come in due time.

"Old Brother Gott, a good man, who has stood the fire in California like a Shadrach, said: 'When I was married and brought my young wife home, she, immediately after supper, set out the family Bible, and requested me to read a lesson and pray. I had never prayed in my life, and I was in a terrible strait. I knew not what to do. I thought it would not do to say *no* to my new wife, so I read, and kneeled down, and tried to pray. I stammered and choked, and made a miserable fist of it; but when she found that I had fairly stalled, she took hold and helped me out. I found that she could pray very well. The next morning I tried it again. Three weeks from that time I experienced the pardon of my sins, and a new heart, and the duty that was before so irksome, now became a delight. The family altar we set up that trying evening was kept up without intermission for thirty years.' Go thou, my friend, and do likewise.

"And you 'ranchers,' (equivalent to members of a bachelor's hall,) who have no families, speak to your companions before you go to bed to-night about this matter. Say to them, 'Boys, we go in for reform. We belong to the Vigilance. Now let us commence to-night and have prayers in the "ranch," and try and reform ourselves.' Don't make a mockery of it, as did a simple-hearted old German, who, having a number of strange guests at his table, said to the one next to him, 'Friend, say grace.' The friend requested the one next to him to say it, and so it passed round till it came back to the old German, and he said: 'Vell, ve can do mitout dis time.' You make the proposition, and then lead the way yourself. The Lord help you. Your eternal happiness or woe may hang upon your action to-night. Will you do it? 'O, but,' says one, 'the boys will laugh at me, and perhaps kick me out of the "ranch."' Men, if rightly approached, are much more considerate and respectful than they get credit for. You cannot know how they will receive such a proposition till you try them. I have no doubt that they will at least behave as well as a lot of gamblers with whom I once had prayers.

"Hungry, wet, cold, and belated, one stormy night, on a trip across the mountains from Santa Cruz to San José Valley, in the winter of 1849, I put up, at a late hour, at an 'old adobe,' which they called a hotel, in Santa Clara. I was conducted into the bar-room, where a jolly set of gamblers were at their cards. After I had taken supper and seated myself by the fire, they got through with their games and profane jokes, and took seats round the fire to look at and listen to the unknown stranger. I gave them an account of things in San Francisco, and especially of the condition of the sick in the City Hospital, which, as some of you remember, was but little more than a charnel-house in those days. Most of those who went there were carried out feet foremost. ['True, true,' said different ones in the audience.] Well, when the proposition was made to retire to bed, I remarked, 'Gentlemen, if there are no objections, we will unite in prayer together before we retire. Let us get down, as some of us used to do when we were little boys with the old folks at home.' They stared at me for a moment in astonishment. The bar-keeper, who was standing behind the bar waiting for a chance to sell to each one another '*nip*' at two bits apiece, said: 'I suppose there are no objections.' So down they got, the last gambler of them, as humble as children, and we had a very gracious season of prayer.

"They then slipped off to bed, as mute as mice. Did they feel like laughing or kicking me out? No, sirs. I met one of them the next day in the town of San José, and he took off his hat before he got within a rod of me. Men will respect you for doing your duty, and whether they respect you or not, *do your duty.* Go home to-night, and try it, and leave the result with God. And you all, each one of you, no matter where you live, or what your relation in life, 'Go into your closet,' or some secret place, to-night, and pray to God, in the name of Jesus Christ, for the pardon of your sins. O, how they have multiplied since you came to California. Would you not like to feel, my friend, that every sin you have ever committed since you were like this little boy, (pointing to a little fellow,) is freely forgiven for the sake of Jesus Christ? Then ask, that you may receive; humble yourself before God, as did the Ninevites; renounce all your sins, outward sins and inward sins, sins of the life and sins of the heart; get an eternal divorce from them all. Will you do it? Will you? 'I have not feeling enough to begin,' says one. Poor fellow, I am sorry for you; your day of grace is almost gone. But try to pray, pray now, pray ever. Pray with the importunity with which M'Donald, the editor of the Sierra Citizen, said he would pray under certain circumstances. Speaking of certain notoriously corrupt officials, said he, 'If I had taken an *oath* to support those men, I would hasten away to the highest summit of the Sierra Nevada mountains, where I would be nearest to the ear of the Almighty, and I would there kneel down, and pray to him to forgive the hell-engendered oath, and if he would not do it, I would *remain* on my knees till the winds should whistle through my fleshless carcass.' That is the determination we want you to have, only in the spirit of Jacob, the wrestler, rather than M'Donald, the editor. But you need not go to the summit of the Sierra Nevada. Jesus Christ is here in the street to-day. He is bending in sympathy over your guilty, blood-bought spirits, *now.* O, speak to him! Reach out the hand of faith and touch the hem of his garment.

"In conclusion, we would give you the advice that a fellow-student of mine from my native place, Rockbridge County, Virginia, gave on one occasion. Said he, at the close of a sermon, 'My unconverted neighbors, I want each of you to pray twice a day in secret, for two weeks, the time of my next appointment here, and then come and tell me the result.' When he returned, a man by the name of Steel, whom we knew well, ran and met him, and told him that he had taken his advice, and though he had no feeling, and could not pray when he first tried, he kept at it, and became so distressed on account of his sins, that he could not do anything else than pray, and that God, for Christ's sake, had pardoned all his sins, and made him happy.'

"Go, each one of you, and try sincerely for yourself. Were I to tell you how to make a thousand dollars, you would jump at the opportunity. Religion will be worth more to you than all the gold of California, and to obtain and exemplify it is the only way you can promote a genuine reform in society. Would to God that all that hear me this day, were not only almost, (as you are,) but altogether such as the Apostle Paul."

CHAPTER LI.

A "LEGION" OF CALIFORNIA DEVILS.

The following exposé of California devils was made on the corner of Sacramento and Liedsdorf streets, in two discourses. The first on Sunday afternoon, the tenth of August, 1856, the second, on the Sunday afternoon following. The congregation in each instance numbered about one thousand hearers. On the second occasion I was honored with the presence of our good Bishop Scott. The text was selected from Mark v, 6–9: "But when he saw Jesus afar off, he ran and worshiped him, and cried with a loud voice, and said, What have I to do with thee, Jesus, thou Son of the most high God? I adjure thee, by God, that thou torment me not. For he said unto him, Come out of the man, thou unclean spirit. And he asked him, What is thy name? And he answered, saying, My name is Legion; for we are many."

"What a variety of devils were contained in the *legion* which possessed the demoniac of Gadara. Every unrenewed heart is the receptacle of 'unclean spirits,' to which they have ingress and egress at pleasure. The difference between the Gadarean demoniac and these California sinners, is that he had more of them at one time than these generally have, although much of the lunacy of the present day is doubtless produced by the same cause. We believe that, with the ripening experience of succeeding centuries, the old 'prince of the power of the air' has effected a more perfect organization of his diabolic forces now than he had eighteen hundred years ago, and, therefore, instead of sending a legion indiscriminately into one poor mortal, he has a great variety of 'bureaus,' or 'departments of state,' with devils specially trained to fill their appropriate offices with honor to his Satanic majesty's government. Some of these grand 'departments,' under the dynasty of darkness, are the following:

"I. The department of Covetousness.

"II. Of Worldly Position and Renown.

"III. Of Politics.

"IV. Of Matrimony.

"V. Of Connubial Infidelity.

"VI. Of Libertinism.

"VII. Department of Slander.

"VIII. The Children's Department.

"And many others, too numerous to mention. We will briefly illustrate the operations of Satan's high officers of state, filling the different departments above enumerated.

"The chief of the *Department of Covetousness* has a splendid 'bazar,' more magnificent than the 'Crystal Palace,' in which is the most gorgeous display of the wealth and splendor of the world. The entrance to this grand palace is free, the doors are always open, and a world of old men, and maidens, and little children crowd in, to see and contemplate the *glory of riches*. They are charged nothing for the sight, but every possible inducement is held out to all to buy a *chance* in the great lottery wheel of fortune. The walls of this great mart are hung with beautiful paintings, and on every pillar various mottoes are presented in large letters, emblazoned in gold; such as, 'Wealth, the key which unlocks every avenue of pleasure.' 'To be rich is to be honored.' ''Tis money makes the mare go.' 'Money is the lever that moves, at once, both Church and State.' In one place is seen a large likeness of an old man delivering a charge to his departing son, as he embarks on the voyage of business life, with this sentence dropping from his lips, 'My son, make money, *honestly* if you can, *but make money*.' In the meantime, the mammon devil is proposing, and making all kinds of bargains with all classes of society in the sale of tickets for the 'great wheel.' Some awful trades are there made. Old men bartering away their honor, and all their hopes of heaven, for one 'chance,' just as their sun of life is setting. Honest young men consenting to be rogues, and multitudes agreeing to tell 'white lies,' and pledging their lives to overreaching and extortion.

"I saw there a temperance man who had been so strict, up to the age of forty years, that he would not allow his wagon to stop in front of a grog-shop. When he came to California, he made a bargain in this mart, the precise terms of which we can only judge of by his subsequent conduct. He opened a restaurant on Clay-street, near Dupont-street, and became a 'gentleman of the *bar*.' How many poor drunkards he manufactured, eternity only can reveal. He was at the business more than two years, and 'drew a blank' every time; and returning to his family, a poor, disappointed dupe, he died on the passage. His body went down into the deep blue sea, and his poor soul, who can fathom the depths of its dreary, downward flight?

"A man who had been a minister in the East, seemed to have made a sort of conditional bargain, for he opened a store in the southern mines on temperance principles, and would sell nothing on the Sabbath. But after a while he left his *back* door ajar, so that particular friends from a distance, who could not readily come during the week, might be accommodated with a pair of boots or a week's provisions. This *paid so well*, that a few subordinate devils, from the 'grand lottery wheel,' easily prevailed on him to leave his *front* door ajar. They argued that as far as the wrong in the case was concerned, it was in *principle* no worse than to keep the back door ajar, and that he could close, or conceal his design when he should see any squeamish Sabbath observers about. That paid still better. Every turn of the 'great wheel' brought him a prize. He was really 'in luck.' He was soon after waited on by another diabolical committee from head quarters, who said to him in effect: 'Now if you will throw open your doors, you will very soon make your "pile." You have given up the principle already, and it is mean to be hypocritical about it. A man ought to be consistent, and if the old fogies, preachers, and croakers come about you, just say: "O, this is California."' So open went the doors. The next proposition was to introduce a bar into the store. He really did not like to do that, but his alliance and traffic with this high department of state were so profitable, that he *feared* to say *no*, and for the sake of additional prizes, secretly *desired* to say *yes*. So in went the bar, with the specific understanding, however, that should he be enriched within a few months, he would close up the whole concern, go home, put on his religious cloak, and do good with his money. The principle underlying this last transaction was, 'The end justifies the means.' His prizes became so numerous and rich, that he had to employ thirty yoke of oxen, all his own, to haul goods for his store, and had, besides, several hundred milk cows, and already saw, in the prospective, 'the cattle on a thousand hills,' all bearing his brand. But suddenly the wheel was reversed. His store was burned. The Indians 'stampeded'. and destroyed his cattle, and in the space of a few short months, he found himself stripped almost as bare as the servants of David, whom Hanun, the Ammonite, shaved and clipped, and he had almost literally to beg his way to 'Jericho,' (Stockton, California,) to await the growth of his beard. He there went to hard work in a legitimate way, and sincerely repented of his sins before the Lord. We heard him relate, substantially, the facts above given.

"But the most heart-sickening sight in this grand palace of Prince Diabolus, the scene over which the angels weep, is to see Columbia's fair daughters, with hearts beating with gleeful emotion and joyous hope, cheeks covered with innocent blushes, as they for the first time meet the gaze of the crowded 'bazar.' Now they look at the pictures, and now at the mottoes, and now watch the evolutions of the great wheel, and see the brilliant prizes and the happy winners, as they bear them away. These lovely girls, unacquainted with the dangers surrounding them, and having, it may be, no faithful counselors, are enchanted. And now, certain things, claiming to be men, with all the beauty of young David, and apparent purity of Joseph, ingratiate themselves into the affections of these unsuspecting maids, and propose to conduct them to the lord of the great wheel, where they can, on easy terms, obtain a ticket and draw a prize. The details of the trade, and its dreadful results to the young ladies (eternally marked and blighted through all time) and to their fond parents and friends, we need not relate. Covetousness, and not amativeness, is the rock on which most of such have been wrecked, the opinion of the world to the contrary notwithstanding.

"The palace of Mammon, to which we have introduced you, is the most extensive of any of the departments of state under Prince Diabolus.

"*The department of worldly position and renown* is next adjoining, with which there is a free communication by an open door.

"The prince of *political devils* is a very important personage in California. No college in the universe can bestow as many degrees and titles as he, and he pledges his honor to the world that he will 'dub' no man with a title, and will nominate no man for office, who has not proved himself a *true man*, who would almost lay down his life to subserve the interests of the *dear people*, if they will vote him into office, and accept of him as their most humble servant. The walls of his grand forum are hung with landscape and rural paintings, representing the highest state of national prosperity; and also a great variety of mottoes, in large letters of gold, such as '*Vox populi, vox Dei*,' etc. In the rear of the forum is a 'bar,' where the very best liquors in the world, of every variety, are displayed, and offered free as air. A free lunch given into the bargain. In a back room adjoining, all primary elections are held. Some of the tallest feats old Lucifer ever accomplished are enacted in this very back room. In an open hall, contiguous to this room, is a trap-door, through which men had been seen to descend in the dusk of the evening, but it was never known, by the *dear people*, for what purpose, until the intermeddling police of the V. C. penetrated the dark vault beneath, and ascertained that that was the place where 'patent ballot-boxes' were manufactured and 'stuffed.' By every election day, the old diabolical political demagogue has all his matters arranged to his mind, taking special care to have 'good judges of election,' and to have the votes of the free people cast in or contiguous to a grog-shop, the statute of the state to the contrary notwithstanding. In every precinct he stations a host of invisible devils, and enough of visible Billy Mulligans and Yankee Sullivans, to conduct with consummate adroitness the preconcerted plans of his political highness.

"As true Californians we must acknowledge our

obligation to him for most of the illustrious line of officers who have held official position in our new state. We do not pretend to insinuate that we have not had, and have not now, many honest and excellent men in office. Such an insinuation would reflect on the wisdom of our political devil. He always wants a small minority of the best of men to grace his adminis tration. The Yankee Sullivan fraternity have well-nigh defeated him in this regard, and have brought his management of affairs into such disrepute that he is by no means sorry that the Vigilance Committee have taken them in hand. They had been so true to his interest in general, that he did not like to dismiss them; moreover, he feared that they might resent it by an exposure of his secret counsels. I think his Satanic majesty is about ready for a transformation, and the introduction of a different form of administration in California, utterly repudiating the old, and denying that he ever had an identity with it. When inquiry is made for him under his old cognomen, he will be like the old colored 'Gumbo, of Virginia.' It is said of him, when the cholera first visited that part of the country, it raged principally among the whites, and Gumbo said: 'De angel ob de Lord take care of de colored people.' By and by it broke out among the colored people, and some wicked fellows went one night to Gumbo's cabin, and seeing him sitting alone by his lamp, they knocked at his door. 'Who dar?'

said Gumbo. 'De angel ob de Lord,' said one of them. 'What you arter?' 'Come arter Gumbo,' said one. 'Pooh!' blowing out his lamp, 'no such nigga here,' replied Gumbo; 'he's been dead dis tree weeks.'

"Do not imagine, my friends, that we would underrate or trifle with the elective franchise of a free people; but we would have the American people dissolve all alliance with demagogues and devils. We would have them acknowledge and fulfill their obligations to God, the author of all our blessings, for the rich inheritance he has given us in

'The land of the free and the home of the brave;'

and for the glorious institutions secured to us by his providence, through the struggles and blood of our fathers. We would have them substitute an enlightened democracy for a licentious rum-ocracy. We would have them shut up every synagogue of Satan, and everywhere repair the altars of God, which have been thrown down, as did Elijah in the days of Ahab and his miserable old wife Jezebel.

"We want to see an altar of prayer in the habitation of every American family, and to see all, from the gray-headed granddaddy down to the little rosy cheeks on which the blush of guilty shame never sat, bow down together, at every rising and setting of the sun, in the sincere worship of the true God. We would not then be troubled with 'Vigilance Committees,' nor the necessity for them. We would then realize and exhibit the fact that 'Righteousness exalteth a nation,' as we now realize to our shame, that 'sin is a reproach to any people.' O, then, we would immediately hail the welcome rising of millennial glory.

"In regard to the *matrimonial devil*, we will only stop to say that he has made very many hasty, ill-advised matches in California, which have resulted in a great deal of domestic strife and scandal, and furnished employment for lawyers in the prosecution of divorce suits. One night, in the winter of 1849, two parties came to my house to be united in marriage. I questioned them closely, and everything seemed to be right. After uniting one pair of them, as I requested the second party to present themselves, the groom arose, but the lady kept her seat. Said she, 'I have concluded not to get married to-night.' So they all left. About an hour later the same lady returned with another man, accompanied by several witnesses, and asked to be united in marriage. I objected, and told them they were too hasty, and that a matter of so grave importance should be well considered. But they declared that they had known each other a year, and had been engaged for a 'long time,' and that the lady's coming with the other fellow to get married was a coquettish freak, and that the man before me, the true lover, having met them on their way home, and threatened the other fellow with a 'good licking,' they had come at once without further needless delay to enter together into the holy bonds of matrimony. Their witnesses testified to the material facts in the case, and they were married.

"I united a couple in 1853, and they were a very interesting looking party, and before three days had elapsed they came back to see if I could not untie 'the knot.' At another time I married a couple that seemed to be above suspicion, and a more beautiful pair I seldom ever witnessed. The man took his bride home, and there he met with a fellow, who, professing to have superior claims to the lady, gave the bridegroom a terrible beating, and took charge of the bride.

"Again, when called to a wedding, I suspected, from some whispering in an adjoining room, that all was not right with the lady candidate for matrimony. So I said to her, 'Have you ever been married, madam?' 'Yes, sir,' was the reply. 'Is your hus band dead?' 'No, sir,' said she. 'What has become of him?' 'He is in the city, sir.' 'Have you been legally divorced?' 'Yes, sir,' said she. 'I must have good evidence of that fact before I can proceed,' said I. The evidence was produced, and they were married. A few months afterward the same woman, and she is a woman of wealth, came to our Bethel parsonage with a certain doctor in your city, to have me unite her in marriage with the doctor. 'What,' said I, 'have you done with Mr. H., your last husband?' Said she, 'He tried to get hold of my money, and I drove him away, and he has gone to the State of Iowa, and I have come to get married to the doctor here.' 'Indeed! Well, madam, I can't marry you; I think now, I have married you once too often already, and that is enough for me.' She importuned, but I would not marry them, so they went off, and I presume, from seeing them often promenade the streets together, that they were married by somebody else.

"These are a few specimens of the exquisite workmanship of the devil of matrimony.

"We next invite attention to the operations of the devil of *connubial infidelity*. The darkest chapter of California history is that in which is recorded the disruption of family ties, and the darkest pages in that chapter of darkness are blackened with details of connubial infidelity. You could imagine that they were written with a quill from the wing of a fallen angel, dipped in ink made of the burning tears of grass-widows and widowers, and of deserted children, who might well envy the lot of hopeless orphanage.

"We will not tax your patience to listen to a full

exposé of the delicate yet deadly operations of this incontinent devil; but we will exhibit details sufficient to put you on your guard against him. He tries to gain an influence over the most of men who come to California without their families. We would not insinuate that he succeeds with all, for many will have nothing to do with him; but his success has been truly extraordinary and alarming, especially with those who have been very successful in business, and have plenty of money and time at their disposal; also with very many who have been unsuccessful, and have been detained beyond the set time to return to their families. If he can lead men into bad company, familiarize them with debauchery, and fill them with rum, they fall an easy prey to his diabolical designs. He gives special attention to wives, more particularly to beautiful and young wives, en route for California, to join their absent husbands. He employs a great variety of means for their ruin; wine, flattery, and bribes, in the form of splendid presents. In this work of desolation and death he enlists government officers, officers of ships, and gallant gentlemen passengers, as may best suit his purpose. A great many ladies, too, not the lowest classes either, but wives of all classes, when they start for California, are persuaded, by some of the many means employed to that end, that when they get to these golden shores they will enter into the immediate possession of a fortune, revel in affluence, and move in the most respectable circles of 'high life.' After bidding adieu to friends on the other shore, they said, 'Good-by to poverty, and toil, and care; I'm going to the land of gold.' Many such arrived just in time to hear of the fortunes their husbands had lost, and of those they came very near making, and of those they yet expect to make, if they have 'good luck.' The poor wife, instead of stepping into a mansion all furnished, is conducted by her overjoyed husband to a small upper room, or rented house, in the city, or to a log-cabin 'on the plains,' or to a shanty in some 'gulch' in the mountains. Everything is different from what she expected. She has to go to work with her hands, a thing she never intended to do again, and that, too, under great disadvantages, in the absence of the facilities she has left behind. She has nothing to supply the loss of all the pleasant home associations from which she has torn herself, except the company of her husband, and he is absent most of his time at work or attending to outside business. She now realizes the wonderful disparity between the visionary ideal and the reality of California life, and feels that in some way her husband is to blame for her unhappy surroundings. She becomes dispirited and petulant, and her husband, not appreciating the great change in her circumstances, and laying to heart some of her fretful sayings, is not sufficiently patient and sympathizing. Now a series of domestic broils commences, followed by occasional partial reconciliations.

"Now the smooth, graceful, honorable Mr. Mustachio, dropping in occasionally to spend an evening, as the special friend of the family, manifests a great deal of sympathy for the poor woman. He thinks it a pity that a lady of so rare excellence should be reduced to such drudgery, and to cheer her up he presents her with a few 'gold specimens,' and some arti cles of jewelry it may be. Then he thinks that, with the consent of the husband, an occasional 'carriage or buggy ride' would be very serviceable to her health; and, as the husband has neither time nor money to spare, and as the *honorable* gentleman has plenty of both, his services, so *disinterestedly* proffered, are very cordially and thankfully accepted. All that is necessary now to restore the good lady to cheerfulness of spirit and a happy reunion of feeling with her husband, is to accompany this generous friend of the family to a few *balls*, and a round or two at the *theaters*. This process, together with the inspiring influence of wine, does up the business for that family. Woe betide the poor children who are the witnesses and victims of such a tragedy; for a tragedy it is, more dreadful than that of carnage and blood. Volumes might be filled with the facts which have come to light demonstrating the truth of what I say.

"One or two cases from real life may suffice our present purpose. Mr. A. and Mr. B. (I have their real names, but need not give them) bought and improved a 'rancho' in co-partnership in the Sacramento Valley. A., who had a wife 'at home,' was an honest, hard-working man. B., a single man, was a shrewd young lawyer, and had been a California legislator. He managed to do the trading and 'outside business,' while A. did the hard work on the farm. He managed, also, to get all the profits, and, finally, the title to all the land, and kept the old man poor. He persuaded A. to send for his wife; thought she would be very useful in the 'dairy,' and also save them much expense for washing bills, and could do their cooking. She came, a fine-looking lady, expecting to settle down with her dear husband in the enjoyment of an easy fortune, on their plantation in the Sacramento Valley. But she soon found, to her great disappointment, that Mr. B., and not Mr. A., was the man of means. She could not work, and must be a lady of leisure, at all hazards. Her husband, though affectionate and kind, was nothing but a poor, honest, hard-working farmer. Mr. B. was a *gentleman*, an honorable legislator, popular, young, and good-looking; and, withal, was much more attentive to her than her husband had time to be. He took her in his buggy, and showed her the beauties of the landscape. In a short time they united in driving the poor old man off the 'ranch,' and together took sole possession. They thus got along delightfully till the next election, when Mr. B. again ran for a seat in the State Legislature; but his friends had dropped him, and he was badly beaten by a man of inferior talents. He then got mad, and left the stolen wife. Mr. A. had groped away in disappointment and despair, his wife took refuge in a brothel in a neighboring city, and the Hon. Mr. B. was justly consigned to infamy. He is now in ——, trying to worm his way into the practice of the law, with very poor success. Such cattle will be 'marked' in the future of California.

"I saw a poor man dying, up country, a year ago, and inquired the cause. Well, he had had an accomplished wife, the landlady of his hotel, and she was a doctoress. His excellent lady took a notion to visit her friends in the East; and at the same time to buy a stock of drugs, and then return to California, and give all her attention to the practice of medicine. To accomplish all this, it was necessary for her to have a certain amount of money, to raise which her kind husband mortgaged his hotel, and everything he had, to their full value, and gave her a splendid outfit. When she bade him adieu, her heart seemed almost to be breaking; and she concealed her face in her handkerchief to hide the tears she *did not shed*. A *gallant young gentleman*, a particular *friend* of the family, who had been boarding at the house for a number of months, was missing about the same time; and the poor husband soon learned that his gallant boarder and his wife had set up for themselves in the City of San Francisco. The 'nice young man,' as usual, soon deserted the runaway wife; and she found—I was going to say a home, but I will not degrade that endearing name with such an association—a hiding-place in a den of infamy. The poor husband, crushed and ruined, sunk beneath the flood.

"Again, on the other hand, very many respectable ladies come here to join their husbands, and find them corrupt and degraded. And, while many unfortunate, yet true and faithful husbands, here, have been cruelly slandered at home, there are many abandoned wretches here, who have long since thrown themselves away; and while living in the deepest degradation and infamy, are keeping up hypocritical professions and promises to their families at home, whom they intend never to see, if they can help it. How often have the wives of such come to California, and wandered in desolation a thousandfold more dreadful than the widow's weeds, through our streets, and valleys, and mountains, inquiring for their husbands. I need not multiply illustrations; you who have been long in California, can supply from the records of your own memory scores of cases. Now, for the accomplishment of all this dreadful detail of desolating evils in domestic life in California, the devil of connubial infidelity has done everything in his power to vitiate correct moral principle in the consciences of men and women. The moral standard on this subject has become so low, and its authority so feeble, that very many have said, in regard to the sacred sanctions of the matrimonial relation, 'Let us burst the bands of God asunder, and cast his cords from us.'

"We will invite your attention to a few potent appliances used by Satan to produce this sad result.

"*First.* The insidious Mormon devil of Polygamy. Six years ago, when their great Western apostle, Parley P. Pratt, was interrogated through a dayly journal in this city, whether or not he really did tolerate and advocate such an enormity, he would not openly avow it, but gave an evasive reply. Now they carry and defend the doctrine of polygamy right into the sacred precincts of respectable Christian families, (such a case came under my own observation but a few weeks ago,) and openly advocate it throughout the land.

"*Second.* The spiritual union doctrine of Andrew Jackson Davis, the sum of which is, the only legitimate and valid matrimonial bond is a spiritual bond ordained by the Creator, from the beginning, or from the birth of each pair, and that the evidences of this union are entire oneness of thought, and feeling, and sympathy, and purpose, and action. But hear what the 'great apostle of modern spiritualism' says on the subject. 'True marriages are natural, inevitable, harmonious, and eternal! By the assistance of interior perception and comprehension, I was enabled to ascertain the glorious and consoling truth, that every spirit is born married! When I gaze upon an infant, a youth, a lonely individual, the voice of intuition and true philosophy says, "That infant, that youth, that lonely individual, has somewhere an eternal companion." Therefore, I perceive and understand that a meeting, and in the present state of society, a legal recognition, of such companions is an outward expression of true marriage. And yet no ceremony, no promise, no written or legalized agreement, can *unite* that which is *internally* and *eternally* joined; nor can these solemnities *unite* that which is *internally* and *eternally* separated.' 'The best evidence that two individuals are not naturally and eternally married is, that by dwelling together they generate discord, discontent, disrespect, and unhappiness.'—*Great Harmonia*, vol. ii, pp. 203, 204.

"It is very easy for all newly-married parties to believe that their union is 'natural, inevitable, harmonious, and eternal,' and therefore valid. But how easy for a man or woman, under the influence of this fallacy, to come to the conclusion, in the moment of any one of the ten thousand petty annoyances in the domestic circle, incident to the rough voyage of life, that a mistake had been made in their union, and that, so far from being bound to each other, they were actually living in adultery, and that it was, therefore, their duty to separate and go in search of the real mate. Who can anticipate the mischief yet to result from such a doctrine? It is boldly asserted that already one million of Americans have become the disciples of Andrew Jackson Davis.

"*Third.* The early legislation of California, exempting men from legal penalties for seduction, has had its influence; and,

"*Fourth.* The very accommodating divorce laws of our state.

"We would respectfully submit a few corrective suggestions.

"*First.* Let the moral bands of society be strengthened: 1. By a careful study and conscientious adherence to Bible teaching on this subject. 2. By a repudiation of those disorganizing forces above enumerated.

"*Second.* Let greater precaution and protection be secured for traveling wives and daughters, en route to California. And let there be an end at once of the separation of families, by emigration to California. If a man cannot bring his family with him, let him remain with them where he is. The exceptions to this rule may be pleaded on their peculiar merits.

"*Third.* While husbands exercise proper restraint over their wives and daughters, in regard to the insinuating advances of corrupt men, in regard to balls and theaters, the recruiting stations of the devil of incontinence, they should, at the same time, and all the time, exercise great patience, and manifest a just appreciation of the change in the circumstances of their families, and the new and unexpected trials to which they are subjected, and always exhibit sincere sympathy, and do their utmost to render home attractive and pleasant. Even the setting of a fruit-tree or a rose-bush will contribute toward a circle of attractions which cluster about the idea of 'home.'

"*Fourth.* Let husbands and wives often talk and pray together, and employ their time and energies, as far as practicable, in their mutual improvement of mind and heart, and in the corporeal, mental, and moral development of their children.

"*Fifth.* Let all adulterers and seducers be marked and shunned as are the seduced; at any rate, until there is indubitable evidence of repentance and reformation.

"*Sixth.* Let the family altar be sanctified and maintained in every family, and let the incense of a pure worship ascend therefrom every evening and morning, the deep impressions of which shall cause every member of the 'family band' to sing down to the evening of their lives.

"'How painfully pleasing the fond recollection
Of youthful connections and innocent joy,
When bless'd with parental advice and affection,
Surrounded with mercy and peace from on high,
I still view the chairs of my father and mother,
The seats of their offspring as ranged on each hand,
And that richest of books, which excels every other,
The family Bible, that lay on the stand.
The old-fashioned Bible, the dear blessed Bible,
The family Bible, that lay on the stand,' etc.

"The Prince Diabolos of libertinism is a very extraordinary spirit, extensively patronized in California. A young man who came to his altars, was required to bring a sacrifice. He was a beautiful youth, in the bloom of health, having a sound physical constitution, a brilliant intellect, a good collegiate education, a competent fortune, honorable parentage, lovely virgin sisters, who almost idolized him, a respectable circle of friends, a bright future of personal opulence and honor, to say nothing of usefulness in the world, and a happy eternity beyond. Well, what will you think, when I tell you that the tyrannical demon required him to bring all the interest he had in all these? You will think that he ought to have fled away as from the 'deadly upas.' But, sirs, he immediately commenced to prepare his offering, health, wealth, his interest in parents and friends, reputation, prospects of future weal, conscience; his interest in the atonement of Jesus, and all his hopes of heaven, all, all were brought and laid on the altar of base passion, at the feet of an infamous woman, and then 'he goeth after her straightway, as an ox goeth to the slaughter, or as a fool to the correction of the stocks, till a dart strikes through his liver.' I heard his dying wail as he sank beneath the dark waters of the most horrible of all deaths. I can take you to our hospitals, within twenty minutes, and show you the wrecks of such youths as they lay 'stranded' along the coast of death.

"*Of the department of slander*, we will simply remark, that the slang devil, who is beginning to figure pretty conspicuously in California, is one of the very meanest of all the catalogue of devils.

"*The children's department* is managed with the most consummate skill. If a contract could be made with his Satanic majesty, by which we could convey to him, by a bill of sale, all California sinners over forty-five years of age, in exchange for all the children under fifteen, it would be a most advantageous arrangement for the Church and the world. But Satan will make no such bargains. He feels that he is sure of you, old sinners. He has already foreclosed a mortgage on the most of your souls. The 'time for redemption' has nearly expired. He has you safe enough, and is determined to have the children too, for which end he employs extraordinary and various means, too numerous to mention. He provides amply for the education of their heels and toes, to the neglect of their heads and hearts. Great pains are taken to prepare young girls for a graceful display in 'polite society,' where the smooth, insinuating Messrs. Mustachio figure so gracefully, with most disgraceful results. Worldliness and vanity seem to be paramount in the training of children, while sobriety, industry, and 'the fear of the Lord, which is the beginning of wisdom,' are secondary considerations. To facilitate this great work, fashionable balls and theaters come in for a full share of the business, and 'Lee and Marshall' give annually a free ticket for all the public schools of our city to their 'circus,' with all the elevating associations of the 'pit.' When the parents of these precious germs of immortality are called to answer for the responsibility involved in their solemn charge, and the Judge Eternal shall say, 'Where are my lambs committed to your care?' how shall they be able to stand?

"Now, in conclusion, allow me to inquire, 'What are you going to do about this diabolical business?' The facts we have educed, substantially, you cannot deny, and they suggest legions of kindred facts to your minds, equally dark and damning. However we may differ in regard to the theories by which to account for these facts, we all agree as to the facts themselves, and that they ought to be remedied. Some of you have denied the *existence of devils*, and others of you have repudiated the doctrine of *human depravity;* but I ask you, How do you explain these extraordinary phenomena, the actual developments of sin, the truth of which you all admit? If you deny the existence of devils, then you have to set down all these deeds of darkness to the account of human depravity, and you thereby make man nothing more nor less than an incarnate devil. If you repudiate human depravity, and set down all these iniquitous developments to the account of Satan, then the question arises, How is it that men, purely innocent, should be so completely under the dominion of the 'wicked one?' The only rational theory is that revealed in the Bible, which asserts the existence of devils and their commerce with men, and asserts as well the deep depravity of the human heart. You cannot account for the dreadful developments of sin in your own hearts, and its manifestations all around us, on any other theory. Now, where shall we find a remedy? The only physician who can effect a cure, is He to whom the demoniac of Gadara applied. Many of you have tried other means, but you have got worse and worse. His terms are, that you come to him, without price, submit your case, and trust him for a cure. See the man from the tombs; he is coming to Jesus. See him run and fall down at his feet, worshiping him, and exclaiming, 'O, Master, for pity sake, save me from these tormentors!'

He is a sincere penitent, depend upon it. What is that he is saying? 'What have I to do with thee, Jesus, thou Son of the most high God? I adjure thee, by God, that thou torment me not.' That is your sincere penitent, is it? 'He is drunk, or crazy, or else he wants to turn the whole thing into ridicule.' Nay, that is the devil that is speaking through the vocal organs of the poor sinner. The man is a sincere penitent, nevertheless. His case exhibits outwardly the secret struggle you feel in your heart to-day. The Spirit of God is working on his conscience. Under its influence he runs to Jesus, and sues for mercy. The devil is also at work with him. Under his influence he cries, 'What have I to do with thee, Jesus, thou Son of the most high God? Depart from me.' Have you not felt this war of the spirits a thousand times? Do you not now feel that you ought to come to Christ? And do you not feel a desire to become a sincere Christian? The Spirit of God is now 'working in you to will and to do his good pleasure.' That is the *attraction* of grace that leads to God. Do you not, also, feel an inward tide of carnal enmity, 'not subject to the law of God,' manifesting itself to your consciousness in spiritual apathy, hardness of heart, insubordination to the will of God, and which says to Jesus, 'What have I to do with thee?' That is the *repulsive power* of diabolized human depravity. Why will you not now come to

Christ and be saved? 'O, my heart is too hard.' There it is, sir; that is the very thing to be remedied, and Jesus Christ alone can remedy it. If it be human depravity, you cannot *free* yourself from it, nor *subdue*, nor *improve* it. You must submit it to the great Physician just as it is. If it be the devil in possession of your heart, do you imagine that you can convert the devil, or free yourself from him, or get his consent to let you come to Christ, without a violent struggle? You can do not one of these any more than you can vail with darkness the noonday sun. If you are ever saved, you have to come to Jesus, as did the Gadarene. Bring your depravity, just as it is. Bring your devils, in spite of all their clamor and opposition, to the feet of Jesus. He is here in Sacramento-street, this afternoon, as really as he was present to the poor man who had the 'legion,' present in his spiritual nature, in his essential Divinity, in all the plenitude of his saving mercy, *now*. He is very desirous to save you from your sins, and from Satan's power to-day. Never say again, 'I cannot go to Jesus, because I do not *feel like it*.' That is the strongest possible argument why you should come without delay.

"Allow me, for your encouragement, to give you a few illustrations of the *attraction* and *repulsion* you feel to-day; that war of the spirits in your bosom.

"John B. Youngs, a friend of mine in Baltimore

City, went to the altar as a seeker of religion, because his judgment was enlightened, and he believed it to be his duty; but for a week, though he went forward every night, his heart was so hard, that, when any brother went to speak to him, he could hardly keep from *cursing* him, and telling him to go away and let him alone: but he submitted his case to Christ, and was cured, and became a most exemplary and useful Christian.

"One Sunday night, in the fall of 1853, after preaching and prayer-meeting in 'the Bethel,' a pale, sickly-looking young man attracted my attention. Taking him by the hand, I said, 'How are you, my friend? you appear to be unwell.' 'Yes,' replied he, 'I am unwell, and am a very miserable man. I have been to the hospital, and am just recovering from a spell of sickness. I am hardly able to work, and have no employment if I were able. I am out of money, and have not one friend in God Almighty's earth.' I then passed round his hat among the brethren, and 'made a raise' for him. I then said to him, 'Would you not, my friend, like to enjoy religion; to give your heart to God, obtain the pardon of all your sins, and become a good Christian?' 'I would, sir,' said he. 'There is nothing in this world I desire so much as religion.' 'Are you willing,' said I, 'to kneel down here, and ask God, for the sake of Jesus Christ, to give you pardon?' 'No, sir,' said he, 'not now.' 'Why not now?' 'I don't want to make a mock of the thing,' replied he. 'My heart is as hard as a stone, and feeling as I do, it would be hypocrisy for me to kneel down there. I don't want to seek religion until I can feel more on the subject. It will do me no good to say a prayer unless I feel it in my heart.' 'Have you been religiously trained in your boyhood?' I inquired. 'Well, sir,' said he, 'when I was a little boy my parents moved from New-Jersey to the wilderness of Indiana. I was there brought up in isolation from all society. I never had a friend. I never saw but one man with whom I sympathized, and don't know that I ever saw one that ever sympathized with me. My father was a professor of religion, but he did not pray in his family, nor do his duty. I saw him die, and did not shed one tear, such was the hardness of my heart then, and I feel just so now toward all the world. I heard you preach this morning, and you struck a tender chord in my heart, and I came back to-night, hoping to receive some benefit; but you didn't strike the right string to-night. And those cold, hypocritical prayers; my God, they like to have killed me. When that man was praying over there, I could hardly keep from telling him to hush.' 'Well, my friend,' said I, 'you are in a bad condition, and the sooner you get your case into the hands of the great Physician, the better for you. In the midst of al this carnal enmity you exhibit, I perceive, by the desires you express, and by the tears you are shedding, (for he was weeping through the whole conversation,) that God's Holy Spirit is operating on your heart, and it will be at the peril of your soul that you leave this house to-night without submitting to the claims of God upon you. After all, it is your pride that prevents you from bowing as a penitent before God, in presence of these brethren.' 'That is a fact,' replied he; 'I believe you are right.' And down he kneeled, and went to praying with all the apparent earnestness of the poor 'publican.' There he wept and pleaded for mercy, under the *attractive* power of the good Spirit, for about thirty minutes, when suddenly, just as in the case of the Gadarene mourner, the wicked spirit got the ascendency, and the *repulsion* was so strong, that he at once ceased to pray, and said, 'My God, I have heard enough of cold preaching and praying in my time, and have seen enough of hypocritical pretensions to damn the world. I wonder that I am out of hell. I don't believe I ever saw but one good Christian in all my life. I sought his acquaintance, but he did not reciprocate my attentions, and I do not certainly know that he was a good man. The Church is full of hypocrites.' 'Well, now, my good fellow,' said I, 'the hypocrisy of the Church, and the sins of other men, are not the questions for you to discuss at this time.

The question is now between God and your own soul. You have sinned, you are under sentence of death, and in bondage to Satan. God now invites you by his Spirit to be reconciled to him. What have you to do now with the sins of others? You have enough to do to dispose of your own, and your only hope is to cast them upon your crucified and risen Jesus.' 'That's true,' said he; 'O Lord, for the sake of Jesus Christ, have mercy on my poor soul.' He then prayed earnestly for about twenty minutes, we laboring with him with appropriate zeal. Then he suddenly stopped again, and said, 'I don't want man to convert me. If God don't convert me, I don't want to be converted at all.' 'Well, then,' said I, 'call upon him; all the men in the world, and all the angels in heaven added, cannot relieve you. Jesus Christ alone is able to save you.' He went at it again with increasing earnestness, and in half an hour was, like David, brought up out of a horrible pit and the miry clay. His feet were set upon a rock, his goings were established, and a new song was put into his mouth, even praise to God. Two years afterward he took his certificate of Church membership as he embarked for the East to see his widowed mother.

"Now, my dear sirs, if you would decide the issue of that war in your hearts, and save your souls, come to Christ *now, just as you are.* Here is Captain M'Donald. Many of you know that he was one of the worst sinners in this city; utterly abandoned. He had given up all hope of ever trying to reform. Three years ago, listening to the preaching of the Gospel on the Plaza, a ray of hope—the hope that, bad as he was, he yet might be saved—penetrated his heart. He went from the Plaza to the Bethel, and before ten o'clock that very night he was happily converted to God. He has walked in and out before you ever since; and put your finger on a single deviation from consistent Christian deportment in his conduct if you can. He had been all through the Mexican war, and for many years in the United States Navy. His associations were all bad, and up to within a few weeks of that blessed day when he surrendered himself to God, he had not drawn a sober breath for ten years. See what the grace of God hath wrought. 'We pray you, in Christ's stead, be ye reconciled to God.' Proceed at once in the light of these examples, and close with his proffered terms of saving mercy."

A HARD CASE.

On Tuesday, the nineteenth of August, 1856, the second day after the above discourse was concluded on Sacramento-street, a very intelligent-looking man called at my house, and requested a private interview. He was trembling with emotion, and seemed afraid to give utterance to what was evidently a great burden on his soul. I tried to make him easy, assuring him that I was willing to do anything for him I could.

Said he, "In your discourse on California devils you gave a number of cases of men who had been guilty of seduction and connubial infidelity, and, though you concealed their names, I know I must be one of the men you referred to. You have doubtless heard of my case."

"No, sir," said I; "I never knew nor heard of you before."

"Well," continued he, "mine is a dreadful case, as bad or worse than any you mentioned in your sermon." He then paused, and groaned, and wept. "I have never told a living man my situation, but my distress of mind was so great I thought I would come and see you. I thought some of going to see Dr. Scott, but I concluded that, as you had been so long in California, you would know better how to treat my case than anybody else. I walked for a quarter of an hour on the sidewalk in front of your house, questioning in my own mind whether to come in or not; but something seemed to say to me, 'Go in; if you do not, you will certainly go to hell.' So I came in, but no living mortal can tell what I suffer; and yet my sufferings are nothing to what I deserve."

Again he paused and wept. Then resuming his

story, he said: "Father Taylor, the confession I am about to make is of the most dreadful and humiliating character. O that I could blot it out forever from the book of memory, and the book of God. In the State of —— I was a minister of the Gospel. I then enjoyed religion, and tried to lead a holy life. I was happy. I left there a pious, good wife and two children, six years ago, and came to California. I engaged in the practice of law, made money, lost my religion, and became an occasional gambler; I seduced a man's wife, and took her away. The man afterward died. I got tired writing lies to my wife at home, and ceased to write, and receiving no letters from her for a year, I concluded she was dead. I then got married to the woman I had taken, and we have ever since lived together and passed for husband and wife. I afterward learned that my wife at home was not dead, but was still living in the same place with her children. I have three children by my present wife. This thing has preyed upon my mind for years. I have often talked to my wife here about it, and she has also suffered great distress on account of it. She is a sensible woman, and wants to be a Christian. In my distress I have resorted to the bottle for comfort, and have been getting worse and worse. I was drinking last Saturday night in the very same house where Captain H. was drinking before he committed suicide; and I feel that I have got down to the lowest point of degradation, and unless I can obtain relief I shall soon be in hell. I should have committed suicide myself long ago, and put an end to my earthly sufferings, but I am afraid to die. I never will in my senses take my own life; for in all my wickedness, I have never doubted the truth of the Bible, and I dread the future. I believe in the atonement by Jesus Christ, and that is my only hope; but my relations in life are so complicated that I cannot seek his mercy, and I see no chance for my poor soul; but if you can see any way of escape, do tell me what to do. My wife at home is a good Christian woman, and was always kind to me, and our children are smart and interesting. I love the woman I am living with, and she loves me, and we could live happily together if we were legally married; but I'll submit to anything in this world to save my soul from this intolerable hell which I suffer."

"Well, sir," I replied, "yours is a very bad case, but I trust not entirely a hopeless one. I see some grounds of hope in the fact that God still preserves your life, and in the fact that you still believe the Bible, and that you suffer so much remorse of conscience for your wrong doing, and especially in the strong desire you have to repent of your sins and seek the pardoning mercy of God, and in the willingness you seem evidently to have to submit to any mortification or self-denial that God may require, that you may save your soul. Your matrimonial relations are exceedingly complicated, and all we can do in regard to them at present is to settle a few facts and principles in the premises, and then wait the developments of a wise and merciful Providence for the issue. In the first place, your wife at home is your only lawful wife, and yet your other woman has claims upon you for support and for the support of your children by her, which you cannot, in justice, disregard. But while you must not allow them to suffer from neglect, it will become necessary for you to vow uncompromising celibacy, if not forever, at least until the whole matter, in the order of Providence, is settled according to law, and the spirit of the Gospel."

"That I will do," said he, "most gladly, and conscientiously."

"Your only hope," I continued, "in the meantime, is to submit your case unreservedly to God, and sue for pardon in the name of Jesus Christ, and trust to the wise Providence of God to open a way of escape. God can do it, and he alone. You need not wait to see the results of his providence in regard to your tangled relationships, before you can experience pardon, if your will is entirely subjugated to his will, so that you will heartily acquiesce in his decisions, whatever they may be. A man of my acquaintance in the State of Virginia, by the name of Beck, invested more than all he was worth in a distillery. Just at that time a camp-meeting was commenced in the neighborhood. He attended the meeting, and the Holy Spirit called him to follow Christ. He hesitated a few minutes, and said to himself: 'If I seek religion I must give up my distillery. If I give that up I will beggar my family. If I do not seek religion I can make a good living for my family, but my soul must go to hell.' He immediately presented himself at the altar, and said: 'Lord, I'll trust my family in thy care, and seek the salvation of my soul. O Lord, I have built a "still house," which I know I must give up before thou wilt pardon my sins, but I want the pardon of my sins to-night, for before to-morrow I may be dead. O Lord, if thou wilt *trust* me, and, for the sake of Jesus Christ, forgive my sins to-night, I will go home to-morrow morning, if spared, and knock every tub to staves, throw out the still, and never make one drop of liquor.' That very night he was redeemed from sin, and I heard him afterward say, in a class-room, after relating his experience, 'God saw my sincerity, and converted my soul on *credit*.' He kept his word with the Lord, to the letter. He destroyed every 'tub,' and converted the building into a mill. I have often seen his still, for he never would sell it, lest it might be used for the purpose of making liquor, and affect his contract with the Lord. Just so, my dear sir, if you honestly and unconditionally submit to the will of God in advance, in view of all possible contingences, God will 'convert you on credit.' You may obtain mercy today, by unreserved submission and simple trust in a living Saviour." We then spent some time in prayer together.

The third time he called to see me, he was professedly, and, I believe, truly happy in God. When he came again, he said:

"I have written to my wife, and have told her all about my dreadful wickedness and wretchedness. I told her that I had so betrayed her confidence and disgraced myself, that I was not worthy of her, and that if she desired to obtain a divorce, I would furnish her all necessary information and evidence; but that if, in view of all the facts in the case, she still preferred me as her husband, I would make provision for my family here and go home. I have done this with the knowledge and consent of my woman here. I leave the matter in the hands of God. I am happy in the love of God, and I believe he will bring to pass that which is right under the circumstances, and I shall be most happy to submit to his will."

CHAPTER LII.

THE SEBASTOPOL OF "OLD NICK."

The city of San Francisco may, with propriety, be regarded as the very Sebastopol of his Satanic majesty. This city, it is true, can exhibit as many church edifices at a greater cost, than any other city of its age in the world. The people of California are justly proverbial for their liberality in giving for charitable and religious purposes. They also treat a man's religious opinions, professions, and efforts, with more respect, probably, than any other new country; and a minister of the Gospel can preach in the open streets of any city or town in California, day or night, without any fear of serious disturbance. Everybody, to be sure, will not stop and listen, but nobody will stop to interfere with him. But, with all these admissions in favor of California in general, and of San Francisco in particular, I believe, nevertheless, that it is, as yet, the hardest country in this world in which to get sinners converted to God. It was a long time before even Christians would believe it possible to have anybody converted in California; and to this day, the Church in this country has not that faith in the omnipotent power of the Gospel, and its perfect adaptation to the wants of the people of California, which is necessary to consistent and successful effort. I now speak of the Church collectively. There are many men here of strong faith, and consistent zeal in the cause of God. In answer to the question, why the successful cultivation of this field is so difficult, I remark, that there are many causes, a few of which I will note. 1. The migratory character of our population. The Christian does not settle down long enough to get acquainted with his neighbors, and by the time he can be sufficiently drilled for efficient service, he is off, and the ranks are left unsupplied, or, at best, supplied with raw recruits. We bring the truth to bear on a sinner's conscience, and feel that, by the help of the Lord, we are leading him right up the mountain, where flows the blood of the all-atoning sacrifice. We look round to see his flowing tears, and hear him shout as he beholds the crucified, but lo! he is gone. He has taken passage to the mines, or to parts unknown, and we see his face no more. Closely related to this difficulty,

I note, 2. The *isolated* condition of society. In all old-settled communities, each member, however humble, is as a link in a chain of association, which runs through the whole community. Cut one link, and it affects the whole chain. But here the links are nearly all separated, and where there is a connection it is generally by *open* links, which can be slipped at pleasure. We see this illustrated on funeral occasions. It is a very familiar sight here to see an unattended hearse moving toward the city of silence. "Who is dead?" "Colonel B.," says the driver. Had the colonel died at home, he would have been followed by a funeral procession a mile long. (The exceptions to this rule are the funerals of Free-masons, Odd Fellows, and other associations, which give due attention to the burial of their dead.) So in other countries; if you succeed in converting a sinner from the error of his way, you can at once avail yourself of his influence and relationships to society, by which you extend your conquests into the territory of the enemy. Through him you reach his parents, his wife, and children, brothers, sisters, and intimate acquaintances; a score of souls saved instrumentally through that one medium. But in California, though the Gospel "battery" may be as powerful as in any other country, still, for want of "conductors," it does not produce results corresponding with its power. Social ties and relationships, and ties of blood, are very important "conductors" for Gospel "electricity." In these we are deficient. California differs, too, from all other *new* countries in this respect. Our other Western States were settled up by a gradual emigration of families. Every family was a nucleus of social life. These readily united with other families, and very soon communities were formed, bound together by strong bonds of social sympathy. But here we have in the space of a very few years, a population of three hundred thousand souls. A state vieing with the great states of our Union, in the development of its physical resources, while even the foundations of its social life have not been permanently laid to this day. The good families interspersed through the state are as deposits of leaven, and their influence is appreciably felt; but there is such a vast disproportion between the leaven and the mass to be leavened, that if the leaven does not "die," it will be a long time before "the whole lump is leavened."

This young giant of the West (the State of California) is very much like the boy, who, at the age of five years, wore his daddy's boots, and whipped his mamma, and then took to the sea, where he has grown up without parental restraint or the refining influence of virtuous female society. He has an extraordinary intellect, is a noble, generous-hearted fellow, but he thinks as he pleases, acts as he thinks, and does not feel that he needs instruction.

He is like Jeremiah's "wild ass, used to the wilderness, that snuffeth up the wind at his pleasure. In his occasion who can turn him away?" Or like Job's unicorn, of which he says, "Will the unicorn be willing to serve thee, or abide by thy crib? Canst thou bind the unicorn with his band in the furrow? Or will he harrow the valleys after thee?" But if these figures are too unseemly, I should say that he is like Job's war-horse, whose "neck was clothed with thunder. Canst thou make him afraid, as a grasshopper? The glory of his nostrils is terrible He paweth in the valley, and rejoiceth in his strength: he goeth on to meet the armed men. He mocketh at fear, and is not affrighted; neither turneth he back from the sword. The quiver rattleth against him, the glittering spear, and the shield. He swalloweth the ground with fierceness and rage: neither believeth he that it is the sound of the trumpet. He saith among the trumpets, Ha! ha! and he smelleth the battle afar off, the thunder of the captains, and the shouting." But we do not despair of our young giant. He is becoming domesticated, and is beginning to attend Church regularly, and we expect to see him converted to God yet, and when converted he will be a "Saul among the prophets," head and shoulders above his neighbors. There are other reasons why it is so difficult to cultivate Immanuel's land in California; but when I penned the caption of this article, it was with the design of introducing a single illustration, drawn from a sort of naval engagement I had, before the walls of his Satanic majesty's fortifications, a few years ago, the account of which is contained in the following extract from my journal:

"*July* 5, 1852.—On the twenty-sixth of May last, I commenced a protracted meeting on the Bethel Ship, which has just closed. At the commencement I had a hand-bill printed after this wise: 'A meeting, to transact business for eternity, will be held on board the Bethel Ship, at the foot of Washington-street, commencing this evening at eight o'clock, and to be continued every evening for ten days.' The bills were neatly printed, presenting a ship under full sail. These were posted all through the city. At the close of this term, I had another bill printed, after the same style, extending the time for thirty days more. Thus we met, and preached, and sung, and prayed for forty nights. The week-night congregations averaged from thirty to forty persons. This may appear to be a very small attendance, yet these were the largest congregations I have ever seen in this city at any church through the week, unless on some extraordinary occasion, of a single night. I have never been able to hold protracted meetings here with much success on that account. The result of the meeting, so far as manifested, is the conversion of about ten souls, and a very gracious revival in the Church. It strikes me now, more forcibly than ever before, that the diabolic enginery by which the thousands of this city are being whirled with fearful rapidity toward the pit, is extraordinarily strong, more than a million horse power, and no ordinary appliances can successfully counteract it. The same amount of effort which has been put forth here, within the last forty days, would, in any other place I ever saw, produce a manifestly glorious result. Our meeting, however, thank the Lord, was a good one, and I am satisfied that much fruit will yet be seen which does not now appear."

This country is a great rendezvous for the representatives of all nations, which, in connection with the fact of its proximity to the Islands of the Pacific and the teeming millions of Asia, constitutes it the greatest missionary field in the world. What St. Peter saw in vision on the housetop of Simon, the tanner, we see now, in fact, here in California, with this difference, that when the "great sheet" was let down on these shores, it was not drawn up again, as when St. Peter saw it; but the attraction of our gold mountains produced such a commotion in the heterogeneous mass contained in it, that the sheet was rent from one end to the other, and out tumbled the whole concern, and every fellow of them grabbed a pick and shovel and went to digging, and here they are to-day. Nor is one of them to be called "common or unclean," but all of them are embraced in the covenant of promise. Every one of them is the purchase of the Saviour's blood, the object of his sympathy and continual intercessions. Let Christians, therefore, everywhere pray for the conversion of California. Let all the appliances of Gospel warfare be furnished and employed against this Sebastopol of Satan. Let it be stormed by the allied forces of King Immanuel, and very soon we will control the whole empire of darkness, and, under the banner of the cross, will march to the conquest of the world.

CHAPTER LIII.

TRIUMPHANT DEATH SCENES.

An extract from my journal may furnish an idea of what poor patients in the City Hospital had to suffer in the year 1849, and early part of 1850. The city paid five dollars per day for the care of each patient; but bedding, and provisions, and medicine, and nurse hire, were all enormously high; and the profits of the contracting physician would be great or small, in proportion to his outlay for those appliances of comfort for the sick. The nurses too, in most cases, were exceedingly reckless. The visiting committee of the city did not go in unceremoniously, at all hours, as I did, and hence saw less of how things were managed. Many of the facilities of comfort now enjoyed in the hospitals here, however, could not then be obtained. But to the extract:

"*Sunday, January* 13, 1850.—Visited the hospital to-day, (after class-meeting, at which more than fifty persons were present, and a glorious season we had.) Two poor fellows whom I visited, and with whom I prayed yesterday evening, were in their coffins

YERBA BEUNA CEMETERY—THE RESTING-PLACE OF 6,000. ADVENTURERS

Another, by the name of Pitenger, a member of the Baptist Church, from New-York, where he had left a clerkship, at a salary of fifteen hundred dollars per year, was very low. He is quite an interesting young man, but confessed with great sorrow his backslidings. He expressed his confidence in Christ, and hope of heaven, but feared that his unfaithfulness would shut him out." I believe Pitenger went to heaven.

S. SWITZER, OF ROXBURY.

"Poor S. Switzer, from R., was dying. He had been very penitent for some weeks, and professed to experience some peace, though not a clear evidence of pardon. He did not think his end so nigh. There was a peculiarly mournful, yet hopeful interest, attached to the case of Brother Switzer. He has been sick several weeks. Says he: 'I lay for whole nights together without anything to wet my parched lips. A mug of tea is set on the shelf, there, for me; but I am too weak to reach it. Here I lie, in my own filth. I have not been taken up, nor has my bed been cleaned for several days; but though separated from my family, and confined in this dreadful place, I am happy, my soul is happy in God. I shall soon be released, and shall suffer no more.' Tears of joyous hope ran down his sunken cheeks, as he discoursed on the glorious prospects of future blessedness which opened before him." I attended his funeral on the fifteenth of January, 1850. He was a member of the Congregational Church. He informed me that he had a wife and three children in Roxbury, of whom he spoke with great affection, commending them to the care of the Lord when he was dying. I have no doubt that he landed safely in the haven of rest, at the right hand of God. May his wife and children be so happy as to meet him on Canaan's coast.

Query: Why should a good man be reduced to such an extremity of destitution and suffering? According to St. Paul, (Romans viii, 10,) though Christ be in us, implying all attainable goodness in this life, the body is nevertheless dead, legally dead, because of sin; still, under the unmitigated sentence, "Dust thou art, and unto dust shalt thou return," this sentence is alike executed upon old men and babes, good men and bad. As for the extreme sufferings of some more than others, there are, doubtless, disciplinary reasons in the Divine administration, to develop certain graces, and the better to prepare the believing soul for an "eternal weight of glory." The rewards of vital piety are purely spiritual; and all, except the foretastes necessary for our encouragement and usefulness here, are to be revealed beyond the swelling flood of Death's dark river. "But if the Spirit of him that raised up Jesus from the dead dwell in you, he that raised up Christ from the dead shall also quicken your mortal bodies, by his Spirit that dwelleth in you."

The resurrection power of Jesus, through which the believing soul is, by the Spirit of God, raised from the death of sin here, will, by the same Spirit, be applied to every essential particle of our bodies, and thus redeem them from all the evil consequences attendant upon the fearful wreck of sin, under which they went down into corruption's deepest sea. "Then shall be brought to pass the saying that is written, Death is swallowed up in victory."

ISAAC JONES AND HIS WIFE MARY.

In the summer of 1850, this estimable pair arrived at San Francisco, from the State of New-York; I believe from Buffalo. They were natives of North Wales. They moved into a little house next door to our little church on the hill; and presenting their letters, immediately identified themselves with the Church, the neglect of which has proved fatal to the spiritual life of many in this land. Brother Jones was a local preacher, and by trade a printer. He wrought here in the office of the "Evening Picayune." He entered into a special agreement with the proprietor of that journal, that he should never be called on to set type or do any work on Sunday. Some weeks afterward, his employer said to him one Saturday night: "Jones, the steamer has just arrived, and we have so much new matter to set up, that I want you to lend a hand with the boys, and set up a few thousand ems to-morrow." "My dear sir," replied Jones, "I am willing to work till twelve o'clock to-night, and commence work again at one o'clock on Monday morning; but you know I told you in the commencement, that it was against my principles to work on Sunday, and we made an agreement to that effect."

"O well, never mind," said the proprietor.

A few weeks passed pleasantly over the God-fearing printer's head, when late one Saturday night his employer said to him again: "Now, Jones, it's no use talking; you see what a quantity of matter we have to set up for the next issue, and a great deal of it must go in type to-morrow. It has to be done, and you may just as well help to do it as for the other boys to do it all. The fact is, I won't have a man about me unless he is willing to work at all times whenever he is needed."

"Well," said Jones, "I shall be very sorry to lose my situation, for it is very expensive living here, and I am dependent on the dayly labor of my hands for the support of my family; but if my continuance in your office and my support depend upon my working on the Sabbath, I'll beg my bread from door to door, or starve, if need be, rather than desecrate God's holy day."

After bustling round among the type stands a while, the proprietor replied, "Well, Jones, you are a good workman and an honest fellow, and I don't want you to leave me."

Jones was never asked again to work on the Lord's day, and kept his place in that office while he lived, and verified in his experience these words of David: "Trust in the Lord and do good, so shalt thou dwell in the land, and verily thou shalt be fed. He felt a great interest in the spiritual welfare of his countrymen in this city, and organized a Bible-class for their benefit. On the third of November, 1850, he had an appointment to preach to them in his own house. His friends assembled to hear him preach, and, to their utter dismay, found him struggling in the chilly grasp of death.

About ten o'clock the night preceding he was seized with the cholera, which did its fatal work in fourteen hours. He said, as he was sinking, "I have a Friend. It is all light about me. I shall soon get home."

His conduct in life and experience in death were just the opposite of those of a poor fellow I visited in the hospital a few months before. He said to me, as he was nearing "the dreary flood:" "I used to enjoy religion, but when I came to California I thought it was necessary to work on Sunday, and do as other people did, or I could not get along. Working seven days in the week and overdoing my strength, is, I believe, what broke me down and brought on this illness; and O, my soul! I have lost my religion. I shall die and meet an angry God."

No language can portray the wretchedness and despair of that poor man.

Mary Jones was, in spirit and in piety, the exact counterpart of her husband. It seemed there was such an affinity between them that they could not long be separated from each other, for two days after the exit of her "dear Isaac" she was laid low by the same fell destroyer, the cholera. About an hour before her death, after a dreadful struggle with the disease, which appeared to convulse every muscle of her frame, she sat up in her bed, and, clapping her hands, shouted, "Glory! glory! glory! I shall soon meet my dear husband, and shall be with my blessed Jesus forever." She then joined me in singing:

"O land of rest, for thee I sigh;
When will the moment come,
When I shall lay my armor by,
And dwell with Christ at home?
O this is not my home,
This world's a wilderness of woe;
This world is not my home," etc.

She then sung two hymns in her mother tongue. I could not understand the sentiment, but could not be mistaken in the spirit in which they were sung. Her voice was strong and clear, and as sweet, I should say, as the melody of the spheres. At that moment her physician, a German, still living in this city, came in to see his dying patient, and there she sat singing as cheerily as a lark. The doctor stood in the doorway through which he was entering, and gazed in utter astonishment till the melody ceased, and said, "Why, how she sings!"

As the tide of life ebbed out, and she was no longer able to shout and sing, she repeated, in soft whispers, "Jesus, Jesus, O my precious Jesus!" She was buried beside her husband in "Yerba Buena Cemetery," to await with him the certain fulfillment of that Divine announcement by the mouth of St. Paul, "If the Spirit of him that raised up Jesus from the dead dwell in you, he that raised up Christ from the dead shall also quicken your mortal bodies by his Spirit that dwelleth in you."

HENRY DUNN,

Came to California about the close of the year 1849. He was but a youth, on which account, and on account of his uncle, Wm. Eades, an old friend of mine in Georgetown, D. C., I felt a great interest in him. When I found him in the hospital he was still able to walk about, and thought he should soon get better. I invited him to come and stop at our house until he should recover. He accepted the invitation, and said he would come in a day or two from that time. When I called again, however, I was astonished to find that he had suddenly been seized with spasms, which had brought him rapidly down to the gates of death. He, however, had some time before embraced religion, and seemed fully prepared for his exit. On Sunday night, the seventeenth of March, 1850, after recovering from a dreadful spasm, he sung, with a clear voice:

> "My suffering time will soon be o'er,
> And I shall sigh and weep no more,
> In that morning, in that morning,
> And we'll all meet together in that morning.
> My ransom'd soul shall soar away,
> To sing God's praise in endless day,
> In that morning, etc.

"I have some friends before me gone,
And I'm resolved to follow on,
In that morning, etc.
They're seated now around the throne,
And looking out for me to come,
In that morning, etc."

On the following Thursday night, March 21, his tide of life ebbed out, and he quietly sunk in the repose of death. He said to me, a few hours before he died, "I am ready for death. I shall go to my blessed Jesus. I want you to write to Uncle William, and tell him I am going to rest, and, though we shall never again meet on earth, I appoint to meet him, and Aunt Martha, and grandma, and all the family in heaven." And again he repeated, "I shall meet Uncle William, and Aunt Martha, and grandma, and all the family in heaven."

C. R. HOYT.

One day in the month of August, 1850, in visiting the desolate sick strangers of the City Hospital, I entered a "ward" in which there were about thirty patients. After speaking to a number of them, I proposed, as was my custom, to sing an appropriate hymn, and pray with them. I had never met with any avowed objections among the sick to such a proposition, but, on this occasion, a man muttered in an under tone, "We don't want any prayers here." He, however, became perfectly quiet before the prayer was over, and afterward confessed his shame.

Rising from my knees, a dying young man beckoned me to his bedside. Laughing and weeping together, he said, "O, I am so glad to see you! It is so cheering in this wicked place to hear a song of Zion, and the voice of prayer." He then gave an account of his religious experience, and said in conclusion: "Here I am, away from my home and my friends, a stranger in a strange land, dying; but the blood of Jesus is sufficient for me. I have no fear of death." "Yes," I replied, "the blood of Jesus Christ, his Son, cleanseth us from all sin."

"O, precious truth," said he. "If that were the only truth revealed in the Bible, it is sufficient. Upon that one truth we could build all our hopes of heaven. That one truth received and applied, would save the world. Happy! happy! happy! Glory be to God, my soul is happy!" Thus C. R. Hoyt, from Ohio, left the world.

MARSHAL B. BROWN.

In the month of February, 1851, I found in a large ward of the City Hospital, then on Pacific-street, crowded with patients, a youth from Vigo County, Indiana. At first he was not inclined to talk much, but after singing and prayer he would have me sit down by his side, and hear his sad tale of sorrow. He said in substance that he had embraced religion some time before he left home. When he started for California, it was with the determination to hold on to his religion, and honor Christ wherever he went. But in crossing the Plains, hearing nothing but profanity, and having many things to try his patience, he gradually slid away, and finally lost his religion.

"O how wretched I am," said he. "Here I must die in this miserable place. Not one friend to soothe my dying pillow. And then be lost forever!" Meantime, streams of bitter tears flowed down his pallid cheeks.

"What did my blessed Jesus do to you," said I, "that you should run away from him and disgrace his cause? Did he not treat you well?"

"O, yes," he replied, "but I have denied him."

"Marshal, Marshal, what a pity that you denied your Lord! What are you going to do about it?"

"O, it is too late now; I can't do anything. I might have saved my soul, but now all is lost!"

"So thought poor Peter, no doubt," I replied, "when he denied his Lord, and behaved so badly in the presence of his enemies. But Jesus, though grieved at his sin, was loving him all the time, and looked after him. That sorrowful, piteous look! Peter could not stand it. 'He went out immediately, and wept bitterly.' Then was Jesus glad, and the angels rejoiced when they saw Peter weeping. Such is the sympathy of Jesus and his holy angels for poor backsliders. The angel at the tomb of Jesus said, 'Go tell the disciples and Peter that he is risen.' And soon after it was proclaimed, 'The Lord is risen indeed, and hath appeared unto Simon.' Peter sincerely repented, and was not Jesus kind to him? Marshal, Jesus feels just so toward you to-day. Do you hate your sins? And will you give them all up?"

"O yes," said he.

"Then receive Jesus in your confidence, and in your heart's affections, *now*. Trust in him as your best friend, and your almighty Saviour, now. His blood is now sufficient to secure your pardon. Will you do it?"

"I'll try."

I shook his hand, saying, "I leave you in the care

of your merciful Saviour; you must submit your whole soul and body to him, and every moment expect salvation through his blood."

When I went again I saw that he was sinking, but the dark clouds had flown, for his wounded heart had been healed and gladdened "by the washing of regeneration and renewing of the Holy Ghost."

I saw him several times afterward before he died, and always found him patient and hopeful, trusting in Jesus. He requested me to write to his friends, (which I did,) and said, "Tell them to prepare to meet me in a better world than this."

Marshal B. Brown soon afterward bid adieu to the hospital, and was conducted, we doubt not, to that healthful clime where the inhabitants never say, "I am sick."

WILLIAM H. STEVENS.

BROTHER STEVENS was from Winnebago County, Illinois, where he had a wife and six children. He was taken down sick at a boarding-house on Clarke's Point, in Broadway. There were no temperance hotels here in those days. Brother Stevens lay in a "bunk," in the second story of the building. This story was all in one room, and the boarders, of almost every name and nation, were there stowed away in tiers of "bunks," as they have them on passenger ships; only in the roughest style. These tiers not only extended round the wall on all sides, but were built up in crib form, with little passages between them, all over the floor. In this most uncomfortable place, Brother Stevens lingered several weeks, and died. The bar-room, underneath him, was the scene of drunken reveling, profane oaths, filthy songs, and midnight brawls. The sick man, on one occasion, offered the landlord ten dollars if he would, for one night, suspend the noise of the bar-room, that he might have a little rest; but quietness could not be bought at any price. He requested that some one should go for a minister, or any other Christian man, to come and see him, but nobody there had time. Never having met with the brother, and knowing nothing of his case, I chanced to preach, one Sunday morning, in the street opposite his window. Hearing the welcome sounds of a song of Zion, he got out of his bunk, and crawled to the window, and there listened for the last time to a preached Gospel. The text on that occasion was, "The night cometh, when no man can work." He wept and praised God at that window, for the unexpected privilege, and crawled back to his bed, wishing some one would tell the preacher that a dying brother would like to see him. A man whom he had hired to nurse him finally found my residence, and informed me of the sick man. In a few minutes I was by his side.

There he lay, calm and composed. After speaking of his experience during his illness, he said, "I have a dear wife and six children in Illinois. I leave them in the care of Jesus. I have for many years been a member of the Methodist Episcopal Church, and have proved the sufficiency of the grace of God in a great variety of trials. I want you to write to my wife, and say to her, 'I die in peace, and go home to heaven. I appoint to meet her and the children there.'" In a few minutes afterward, without a groan or a struggle, he fell asleep in Christ.

The next day, Sunday, March 3, 1850, I stood on a pile of lumber in Happy Valley, and preached his funeral sermon to a large, attentive audience, in the open air. The text was, "All flesh is as grass, and all the glory of man as the flower of grass. The grass withereth, and the flower thereof fadeth away, but the word of the Lord endureth forever."

ORLANDO GALE.

THE subject of this notice had been sick at the City Hospital for many weeks, with chronic diarrhea, a disease very prevalent in California in those days, and fatal to thousands of early California adventurers.

About the first of March, 1850, Orlando, a mere walking skeleton, came from the hospital to our house, and said that if he remained in the hospital, he knew he would surely die, and wanted to know if I could do anything for him. I told him to make our house his home, and we would do what we could to make him comfortable. Dr. May kindly gave his services as his physician, and for a fortnight I thought he would rally, but he took an unfavorable turn, and went down. On Friday, the twenty-second of March, while Mrs. Taylor was directing his fading eyes to the cross of Jesus, he suddenly found peace in believing, and shouted the praise of God.

I extract from my diary, of March the twenty-fifth, the following notice:

"Poor Orlando died this afternoon. Since he professed religion, last Friday, he has been very peaceful and happy. He leaves, in Lowell, Massachusetts, a widowed mother. I am very glad we took him to our house. He, by possibility, might not otherwise have been converted."

A. C. CHIPPELL.

On New-Year's Day, 1852, I visited the State Marine Hospital, to pay "the compliments of the season" to the sick and dying. After singing and praying in a certain ward, an old gentleman, greatly emaciated, beckoned me to him. He grasped my hand, and wept some time before he could speak, and then said, "O, that precious hymn you sing, 'A home in heaven.' It fills my soul with rapture. Glory be to God! My soul is full of glory!" He made me sing it again, and said, "It makes me forget all my sorrows."

He then told me that he had been a servant of God for fifteen years, and that now, in his extremity of poverty and affliction, God was unspeakably precious to his soul. Said he, "I have a large family in Connecticut, and it would greatly gratify my heart to see them once more; but God, my Father, knows what is best. I do not wish to decide the question, whether I shall get well, or depart and be with Jesus, which is far better. I leave it all in the hands of God. I am ready, and will wait his pleasure. Glory be to God,

"'Not a doubt doth arise, to darken my skies,
Or hide for a moment my Lord from mine eyes.'"

He said to me again, a few days later, "O, I love my friends, and should delight to see them, and to kiss them," the tears standing in the furrows of his sunken cheeks, "but I would not turn my hand to live, unless it is the will of the Lord. I am happy, happy, in God. And," continued he,

"'Jesus can make a dying bed
Feel soft as downy pillows are,
While on his breast I lean my head,
And breathe my life out sweetly there.'"

Such were the dying triumphs of A. C. Chippell He soon afterward exchanged a miserable berth in the hospital for a mansion in heaven, fitted up for his reception by the King of Glory himself.

MORTON, OF ILLINOIS.

I HAVE forgotten the Christian name of Brother Morton, but I have not a doubt that his "name is written in the Lamb's book of life." He was a tried Christian. I saw him frequently during his protracted illness in the hospital, and always found him patient, resigned, and happy.

"All is well," was his favorite hymn, and he invariably joined me in singing it. He requested me the day before his death to write his wife, and tell her, "I die in the faith. I am going to rest. Tell her, I shall soon meet our two dear children who have gone to heaven, and I want her and our two remaining children to meet me in heaven. Tell her, that my life in California is the best end of it."

"What!" said I, "Brother Morton, have you not been sick ever since you came to California?"

"Yes," he replied, "I came here sick, and have been sick ever since, and have suffered everything but death; but Jesus has been so precious to my soul, that it has been the best part of my life."

His was the experience of a weary pilgrim, who had been homeward bound for many long years, and was just catching the joyful recognitions of kindred faces not seen for half a century. The experience of the mariner who had passed through shipwrecks, sickness in foreign hospitals, life among the cannibals, and dreary years in frozen seas; a Sir John Franklin returning, and now he is in sight of home, his long-sought home.

Near this triumphant Christian lay poor Y., from New-York City, who said to me, "My wife is a good Christian woman, but I have lived a skeptic, and a wicked man. Now my skepticism is all gone. O! if I could but send word to my wife that I have obtained religion, and am ready to die, then I should be satisfied. But, alas! I have no such word to send. I am a poor sinner, unforgiven."

Poor fellow, I felt great solicitude on his behalf, and prayed for him frequently, but fear he died without Christ.

On the twenty-first of March, 1850, I took Brother H. with us to witness the dying triumphs of Brother Morton. When I entered his room his eyes were set back and motionless, and I thought he was dead, but upon examination found that his pulsations, though feeble, had not ceased. I supposed him unconscious of anything I said, and yet I thought his favorite hymn might soothe him in his passage through the chilling flood. So I sung,

"What's this that steals, that steals upon my frame?
Is it death? is it death?
That soon will quench, will quench this vital flame?
Is it death? is it death?" etc.

When I commenced singing the second verse, I observed his bosom heave, and then he began to sigh with emotion. On the third line his eyes flashed with heavenly joy, and on the fourth he raised his voice distinctly, and sung,

"To hide my Saviour from my eyes:
I soon shall mount the upper skies.
All is well! all is well!"

As I continued the song, his strength failing him, he responded to the sentiment by oft repeating, "Halleluiah! halleluiah!" And the ecstatic halleluiah died on his lips, when the tide of life sunk too low for utterance. But I doubt not, that on that very day, his blood-washed spirit resumed the theme in sweeter strains beyond the "swelling flood."

JAMES F. DIXON.

This is the Louisiana man referred to on page 71. He had been religiously educated, had a Methodist wife, and was quite an intelligent man, but, alas! had lived "without God." I spent much time in trying to lead him to Christ during his long-continued illness. At one time he said, "I have spent my life in sin, and it is so presumptuous now, when dying, to offer myself to God, I cannot have the face to do it. I cannot think it possible for him to pardon me now. When I was crossing the mountains on my way to California, I got into great trouble in a certain mountain pass; I went away and prayed that God would deliver me, and enable me to get to California; I felt, though a sinner, that God did regard my prayer. But I have no faith in death-bed repentance. It's no use." I urged him to throw himself on the atonement of Christ as the last plank; sink or swim, to hold on to that, and cry to God for help.

This conversation took place on the fifth day of April, 1850. I continued, at different times, to labor with him until the twentieth of the same month, when he said:

"I give up, I give up all to God. O, that I had done it twenty years ago. O, that I could but live to serve him! but to expect his mercy when I am worn out and dying; how can he forgive me?"

He continued to struggle for about a week from that time, when God, for Christ's sake, gave him the evidence of pardon. At his request, I entered his name on the class book as a candidate for Church membership, but he had served a "probation" of only a few weeks, when, by the great mercy of God in Christ, I believe he was "admitted into full membership" in the Church triumphant on high.

C. W. BRADLEY.

Some time in the spring of 1850, on entering a small basement room in the City Hospital, on Clay-street, I discovered, by the dim light of a small lamp, three men, or rather wrecks of men, for they were worn down by disease to mere skeletons. One of them, a tall Frenchman, whose language I did not understand, performed a successful pantomime, to warn me against vermin. I was exceedingly sorry that I could not reciprocate his kindness by a message of mercy from our common Saviour. He, however, seemed very devout while I was engaged in prayer. In another corner lay C. W. Bradley, of Louisiana. He was too far gone to talk much, but gave his name and the address of his wife, and said he enjoyed peace with God, through Jesus Christ, and was ready to die. The last words I heard him utter were, "Tell my dear wife to meet me in heaven." Truly, the rewards of virtue are not in this life.

ROMEO DORWIN.

In my weekly visits to the City Hospital, I found, on one occasion, in the spring of 1851, a great many men who had been scalded and mangled by a steamboat explosion on the Sacramento River. Here lay some with broken limbs, and there a tier of poor fellows with all the skin scalded off their hands, arms, and faces, as a substitute for which they were coated over with some kind of dough. Humanity was so mutilated and masked, that a stranger would have asked, "What kind of animals are these?" Who can tell what those poor men suffered? Five years have since elapsed, but the sighs and groans of that hour still ring in my ears.

In the midst of several hundreds who appeared to have no experimental acquaintance with God, it was indeed refreshing to fall in with such a patient as Romeo Dorwin. He told me that he was from Vermont, where he had a wife and two children; that he had been in the service of God for many years, and though separated from his family (except one son, a young man who was with him) and confined in the hospital, amid such very unpleasant surroundings, still God was very gracious to him. His experience exemplified the truth of a precious announcement of Isaiah: "Thou wilt keep him in perfect peace whose mind is stayed on thee, because he trusteth in thee." He took great delight in talking about Jesus, and in singing his praise, for though far gone in consumption, his voice was clear and distinct, both to talk and to sing. I saw him often during his decline, and was always asked to sing his favorite hymn, the dying sentiments of Bishop M'Kendree:

> "What's this that steals, that steals upon my frame,
> Is it death? Is it death?
> That soon shall quench, shall quench this vital flame,
> Is it death? Is it death?
> If this be death, I soon shall be
> From every pain and sorrow free,
> I shall the King of Glory see,
> All is well! all is well!" etc.

He joined in the song with remarkable spirit and sweetness of voice. It was no ordinary privilege to witness the triumphs of faith over the depressing facts characterizing the experience of a husband and father so far from home. So unable to help his family, so dependent himself; the sights he saw, the sighs he heard, all combining to cast, as it were, the pall of death over the soul and body of any doubtful or doubting man, yet I never heard a murmur or complaint from the lips of Brother Dorwin.

"It's all right," said he; "my Father knows what is best. I would like to see my family if agreeable to his will, but I have given my dear wife and children all up to him; I leave them in his care. I know he careth for them, and he can provide for them with or without me. Though I see their faces no more on earth, I expect to meet them in heaven. Glory be to God! I shall soon get there, and sorrow and suffer no more.

"'Not a doubt doth arise to darken the skies,
Or hide for a moment my Lord from mine eyes.'"

The night he died, which was the fifteenth of April, he seemed conscious for several hours before that he never would again see the sun by mortal vision, and such calmness and composure as he manifested, I presume never was felt by any man leaving home for California. Some of his expressions were as follows: "God is here; he is in this room now. The angels are here, they are now hovering over me; they wait to bear me up to glory. I shall get home to-night."

Asking for a small looking-glass, and taking it into his own hand, he took a last look at himself, seeming to scrutinize his dying features with great interest, and then, laying the glass aside, he resumed his theme of praise to God, and so continued till slowly and softly he fell asleep in the arms of Jesus. Will his family live for God, and meet him in heaven?

WILLIAM CROCKETT.

I BECAME acquainted with the subject of this notice in the month of December, 1850. He was surrounded by the wounded, diseased, and dying, from almost every clime, whom the stern law of necessity had driven to the City Hospital, then on Pacific-street, San Francisco. A city hospital is a most undesirable place, but a gracious refuge for the unfortunate sick stranger. Board and lodging, doctor's bills and nurse's wages, were so enormously high here in those days that it mattered not how much money a man had, if he were sick long, his purse could not but be drained. And after a poor patient "had suffered many things of many physicians, and had spent all that he had, and was nothing better, but rather grew worse," then his only alternative was to go to the hospital.

Young Crockett was far gone with bronchitis when I first found him. He was greatly distressed about the state of his soul. Said he: "I am not prepared to die. I was brought up to fear God and to pray; but when I grew to manhood, I went from my home in Nashville, Tennessee, to New-Orleans. There I fell into bad company, and became very wicked; but all the time, though I could put on pleasantry and look a man in the face, I was tormented by a guilty conscience. My mother seemed to be talking to me all the time in my heart, and pleading and praying that I should give my heart to God. I had no peace day or night, and now I am completely miserable. I know not what to do. O, if I only had religion! I would give the world if I had it, if I could obtain the pardon of my sins."

I assured him that if the world were his, he could not with its price atone for the smallest sin he had ever committed; that it was a dreadful thing to sin against God, involving terrible consequences to soul and body here and hereafter.

"My dear brother," continued I, "what a great pity it is that you have spent your youth in this miserable business of sinning against God. What have you to show for it? But let us see what God says about it. 'As I live, saith the Lord God, I have no pleasure in the death of the wicked; but that the wicked turn from his way and live. Turn ye, turn ye from your evil ways; for why will ye die?' You have willfully departed from God, and there must be an unreserved submission of your will to him; an entire acquiescence on your part in his plan of saving you. Whatever, therefore, is opposed to the will of God must be promptly renounced. What ever he requires must be honestly complied with Do you now hate your sins?"

"O yes, and hate myself for having sinned," he replied, with great emotion.

"Are you willing now to renounce them all, sins of the life and sins of the heart, to give them all up forever?"

"O yes," said he; "I am more than willing."

"Thank the Lord," I added; "he is now working in you to will and to do his good pleasure. His pleasure is to have you 'turn and live.' Do you *now* give up all your sins?"

"I am trying. If I know my heart, I do."

"Do you now consecrate your soul and body to God, 'living or dying, to be the Lord's' without reserve?"

"I do; I give him all."

"You are then ready to receive Jesus Christ as

your Saviour now? His blood now atones for every sin you have ever committed. You may now obtain mercy on his account. His *credit* never was questioned in the kingdom of grace, but he has actually deposited into the treasury of immutable justice the price of your redemption.

"'Believe in him who died for thee;
And sure as he hath died,
Thy debt is paid, thy soul is free,
And thou art justified.'"

He seemed much comforted in believing that Jesus had made a deposite for him, which he might draw in time to save him from eternal ruin. I left him praying and trusting in Jesus. When I called again, I found him peaceful and happy. He said that God, for Christ's sake, had pardoned all his sins, and that he was not afraid to die. He lingered several weeks afterward, quietly reclining on the arm of Jesus.

He was removed from the large crowded ward into a small room alone, where he could pray and praise without interruption. At his request I administered to him the sacrament of the Lord's Supper, after which he seemed almost overcome with joyful emotion.

He said he wanted to join the Church; and that as his mother was a member of the Presbyterian

Church, he thought it would be gratifying to her for him to join that Church, which he would do, if I had no objections. I advised him to do so; and accordingly introduced to him the Rev. Albert Williams, then pastor of the First Presbyterian Church in this city, who took his name. I was soon after called to attend his funeral; and he was decently interred in the "Yerbo Buano Cemetery," in this city.

DYING MESSAGE OF EDWARD MOW.

I THINK it was near the close of the year 1850, I held my ear close to the lips of Edward Mow, who said: "I want you to write to John T. Cromwell, Ontario County, New-York. Tell him that I have been sick ever since I landed in California. I got a little better, and started for home; got as far as San Francisco, and was taken worse. My hope is firm in the Lord, that if we never meet on earth, we may meet in heaven."

He spoke very affectionately of his sister Margaret, and added: "Tell Mr. Cromwell to remember me to all my young friends. Tell him to talk to my sister Margaret." He was calm and peaceful, and quietly sunk into the sleep of death.

BROTHER GUY, AND HIS TWO DYING REQUESTS.

I HAD frequently conversed and prayed with Brother Guy during his illness in the City Hospital; and it was a real pleasure to witness his patience, and to hear him talk of the love of Christ.

A few minutes before his death he said to me, "I have two requests to make of you: Don't let them bury me until to-morrow; I would not like to be buried alive." (They generally buried their dead immediately.) "And I want them to put me down deep enough. I don't want the dogs or cayotes" (a small species of wolf whose howling could frequently be heard from the city) "to get my body."

"Have you no requests to make in regard to your soul?" inquired I.

"O, no!" said he, "Jesus will take care of my soul."

Thus, on the seventeenth day of March, 1850, he closed his eyes on scenes of suffering and woe in the hospital, to open them in the bright visions of eternal day.

CHARLES AUROM.

He was by trade a painter, a member of our Church, and a consistent Christian in life, so far as I know. He was seized with cholera about the first of November, 1850, and was buried on the third. I spent some time with him, on two different occasions, during his short illness. In the intervals between those awful paroxysms, which so quickly loosen every pin in the tabernacle of the soul, he related to me his Christian experience: He had been clearly converted in Philadelphia, whence he came; had enjoyed much of the love of God; expressed with great regret his unfaithfulness in the cause of Christ; but had never since his profession of religion turned his back upon him, and Jesus had never forsaken him; and now that he was dying, he felt his sweet pardoning love and sanctifying grace flowing into his heart. He was very peaceful, yea, triumphant. He was happily disappointed to find that death, so dreadful in the prospective, had lost all its terrors. I thought it well, indeed, that he had an honest, efficient Comforter in the person of the Holy Ghost; for human helps, even sincerely rendered, are of no avail in such a case. I had there an example of those deceptious offers of comfort too frequently administered to the dying, to their own hurt. As I was entering the room of my sick brother, a man met me at the door, saying, "He is nearly gone; he can live but a few hours."

While I was conversing with the patient, the same man came in and said, "Keep your spirits up, Charley; you'll get better directly; you'll be all right in a day or two."

But, happily for "Charley," he had his "spirits up," buoyant with immortal hope. He knew that he would be "all right in a day or two;" for he should be in that country where the people never die.

I delivered an address on the occasion of his funeral, from the door of the house in which he died, to a large audience, which stood with hats off, in Montgomery-street. I laid his mortal remains in "Yerba Buena Cemetery," to await the awaking peals of the last trumpet.

SAMUEL M. RAMSON.

On the sixth of November, 1850, I was called to the cholera hospital, to attend the funeral of Samuel M. Ramson, of New-York. What a scene was there! Some convulsed and writhing in pain, such as the cholera alone can produce. Others had passed the mortal struggle. The monster had done his fatal work, and his collapsed victims lay along the shore of Death's dark river for hours, without any pain, till the swelling tide swept them away.

I talked to as many as could converse, and labored especially for one poor fellow from Connecticut. He said he believed in experimental religion, but did not possess it, and had prayed but a few times in his life. I tried to instruct him, and urged him, with all the earnestness I could command, to try then to pray and give his heart to God. I kneeled by his bedside, and prayed for him, and after bringing to bear upon him all the moral suasion I could call into requisition, I said: "Now, my dear brother, will you make an effort to give your heart to Jesus, and seek the pardon of your sins?"

He replied, very coolly, "I'll think about it."

Poor man, he had had his lifetime to think about it, and had done nothing more. Procrastination was a leading principle of life with him, and it was the strong, all-controlling principle in death.

> "Procrastination is the thief of time:
> Inch by inch it steals till all is gone,
> And to the mercy of a moment leaves
> The vast concerns of an eternal state."

And that precious moment, on which the soul's eternal destiny hangs, is not treasured, but decoyed and stolen, by the same dreadful rogue that stole the rest.

So when this poor fellow was dying, like too many, alas! he asked time to consider, whether or not he would engage in the business for which our whole life was designed.

Not so with young Ramson. His friend told me, that he (Ramson) had embraced religion on his passage to California, and while in the mines, went alone, dayly, among the bushes to pray, and had, up to the day of death, conducted himself with Christian propriety.

THE BROTHER WHO DEPARTED WITHOUT THE SACRAMENT.

On the twenty-seventh of December, 1852, as I passed from ward to ward, in the City Hospital, trying to administer comfort to the sick strangers there congregated, I saw, in one corner of a small room, a man about forty years of age, whose countenance at once engaged my attention. He evidently was very sick, but his face exhibited an air of serenity and pleasantness very rare in such a place. By some singular oversight, though I penned the facts I here relate, and have a vivid recollection of his face, I omitted to record his name. Thank the Lord that the Holy Spirit is not subject to such mistakes in his records in the *Book of Life.* The good brother informed me that he had been a merchant in Baltimore City, where his family was then residing, or perhaps a short distance out of the city. His lungs were affected, but he hoped it might be the pleasure of the Lord to spare him for his family, of whom he spoke with tenderest affection. He informed me that he was religious at home, and still enjoyed a sweet sense of God's pardoning mercy; that religion made his soul happy when he had nothing else to comfort him. Said he, "On your next visit to the hospital I want you to administer the sacrament of the Lord's Supper to me. I have not enjoyed the precious privilege of the sacrament for a long time." I told him I would do so with pleasure, and should have made preparation for it at once, but he seemed to be in no immediate danger, and preferred waiting. The next day, as I was informed, owing to the breaking of an abscess on his lungs, he suddenly died. He was not permitted again to "drink the fruit of the vine" in commemoration of his bleeding Saviour, but he went, we doubt not, to "drink it new with Christ in his Father's kingdom."

THE DYING GERMAN IN A STABLE.

During the autumn of 1853, the City Hospital, which could accommodate about three hundred patients, was crowded to excess. One day, after visiting the wards of the sick and dying, and passing out into the back yard, a nurse said to me: "There is a very sick man in the stable," pointing to the door. I entered, and saw the emaciated frame of a tall, intelligent looking young German. He told me he was a druggist, had been well brought up, and was doing a good business in his father-land, when he took a notion he would come to California. He had been at work in the mines, and got his leg broken. It had been too long neglected, and mortification had taken place, and he feared he never would again see his dear mother. I explained to him our guilty and exposed condition as sinners, and told him of a Friend, his Friend, one who loved him more than his mother ever did or could love him; that his mother in Germany knew not the condition of her son, and could not help him if she did; that this Friend knew all about him, and that he was nigh at hand; that he was born in a stable, and was present then, in his spiritual nature, his essential Divinity, in that mean stable, and was waiting to receive him as his child.

Never before did I see a poor soul drink in the simple Gospel with such avidity. His faith seemed to follow me closely step by step, till, by the mercy of the Lord, I led him to the cross. Gazing with wonder, he at once recognized the dying Jesus as the victim slain for him. His faith took right hold of the atonement and exulted in an almighty Saviour. His countenance shone like that of Moses, as he ex claimed, " O, my Jesus, my Jesus, I do love thee!"

As he continued to praise God he, every now and then, turned his beaming eyes toward me, and said, "I am so glad you came in to see me. I did not know Jesus till you came in and told me about my precious Saviour. I would like to get well, that I might do something great for you."

I assured him that I was repaid a thousandfold in seeing him happy in God. His strength failing, he said, " My poor pody, he is very sick, he will soon go down; but my spirit, he is well now, he will soon go up to my blessed Jesus." A few hours sufficed to end the mortal strife, and his spirit went up to his blessed Jesus.

THE DYING NORWEGIAN BOY.

In the spring of 1853 there lay in the State Marine Hospital of this city, a Norwegian boy, nigh unto death. Peter Johnson, a Swede, who had been incurably mangled in the mines, and who had experienced religion during his confinement in the hospital, occupied the next cot. Peter spoke very good English, but the boy could only speak his mother tongue. A priest passing through the ward, requested the Swede to be his interpreter, while he should enlighten the dying boy at his side.

I have always found the priests of the Romish Church very regular in their visits to the hospitals. The following conversation in substance was held between the said priest and the Norwegian youth, through the interpreter:

"My boy," said the priest, "you are dying. I am very sorry to tell you that if you die in your present state, you will certainly go to hell. Now, my dear boy, if you will confess your sins to me, and give your soul into my care, I will get a pardon for you, so that when you die you may escape the pains of hell and get to heaven."

"I cannot trust you," replied the boy; "my fader and mudder taught me to confess my sins to my Got in heaven. I believe my fader and mudder; I don't believe you. I cannot trust you, but I will trust in my Got in heaven."

The priest immediately left, and the boy soon after died, "trusting in his Got in heaven."

Parents, sow pure Gospel seed in the virgin soil of your children's hearts. They will need your teachings to guard and guide them when you are dead.

JOSEPH M. GUSTIN.

I BELIEVE Brother Gustin was from Pennsylvania. He passed very peacefully through the scene of starvation and suffering in the "Ex-City Hospital," described on page 66. He died about the middle of March, 1850.

A short time before his final adieu to the gloomy hospital scene surrounding him, he told me that he had "enjoyed religion for three years." In conversation with him, I quoted the seventh verse of the first chapter of the first Epistle of John: "But if we walk in the light, as he is in the light, we have fellowship one with another, and the blood of Jesus Christ his Son cleanseth us from all sin." He replied with much feeling, "O how sweet that is."

Peace to the ashes of Brother Gustin.

ISAAC ENSLOW.

Brother Enslow was from New-York City, by trade a sail-maker. He was a tall, noble-looking man, and had in him a very generous heart. I never heard any complaints from any source against the Christian character of Brother Enslow, during his sojourn of two or three years in California; but he was not known as a very active member of the Church, till our forty days' meeting in the Bethel, in the spring of 1852. At that meeting he took hold like a man thoroughly furnished, and did efficient service. About two months before his last illness, he professed to be wholly sanctified to the Lord, and his conduct corresponded with his profession to the last. About the fifth of August, 1852, he was taken sick with a fever. He was most peaceful and triumphant on his bed of death. He spoke with great affection of his wife and daughter at home, and he would be delighted again to see them; but, said he, "Jesus is my Saviour and my King. He saves me now from all sin, and he has a right to rule. I most gladly submit to the decisions of his infinite wisdom. He doeth all things well. My soul is full of light and love."

I bade him adieu to attend a camp-meeting a

couple of hundred miles north of this city, hoping to find him convalescing on my return, but I saw his face no more. While I was laboring at the camp-meeting he passed from labor to reward. His dust slumbers in the "Yerba Buena Cemetery," to await the resurrection summons of Him in whom he trusted. His death occurred about the fifteenth of August, 1852.

JUDSON FORBES.

On the second day of June, 1856, I was called to the United States Marine Hospital to see a dying man, who had requested an interview with me. Kneeling by his side, I asked, "Do you know me?"

"Yes, Father Taylor."

"Well, my dear friend, what can I do for you?"

"O, I want to know what I must do to be saved. I am a great sinner. I know not what I shall do."

"Have you long felt yourself to be a great sinner?"

"Ever since I left home, a mere boy, and went to sea, my mother's prayers have been ringing in my ears. She used to pray with me every night. I have often thought what a dreadful thing it is to be a sinner, after having the instructions of such a mother. I have often desired to have religion, but at sea I had poor opportunities, and did not know how to obtain it. For some months past I have been in great distress of mind, but I have had no one to teach me the way. I have been trying to pray, but I get worse and worse. For several days past I have felt such a load of sin that it seems sometimes that my heart would burst."

"Do you think," said I, "that you hate your sins, not only on account of their consequences to you, but because they were perpetrated against a wise and merciful God."

'Yes," he replied, "and I give up all to him."

' Even if you knew you would get well," continued I, "you would consecrate your heart and life wholly to him, and living or dying, be the Lord's without reserve?"

"Yes," said he, "I only desire to live that I may serve him."

"Do you believe that Jesus Christ died to redeem you from sin?"

"I do."

"Do you believe that God the Father accepts the price which Jesus paid for you?"

"I do."

"You believe then, though guilty, bankrupt, and condemned, that on Christ's account you may to-day obtain the pardon of all your sins?"

"Yes, I do."

"Are you not glad that you have such an almighty and sympathizing Saviour on whom you may cast all your sins and sorrows?"

"I am glad."

"You are trusting in him now, are you not?"

"Yes;" and continued, "O! O! O!" the tears streaming down his sunken cheeks, "I never felt so before. I feel such a load taken off my heart. I feel that I could fly away to the arms of my blessed Jesus. I never before had any idea of the ability and willingness of Jesus to save me. I feel that he hath saved me. He hath cleaned me through and thro' gh. I hope to see my God in heaven before to-morrow morning."

He, however, survived two days longer, and sweetly fell asleep, trusting in an almighty Saviour. Such was the closing scene of Judson Forbes, from Wisconsin, I believe he told me; aged twenty-four years.

THE END.

SEVEN YEARS'
Street Preaching in San Francisco,

EMBRACING

INCIDENTS AND TRIUMPHANT DEATH SCENES.

TESTIMONY OF THE PRESS.

"AMONG the first of our noble army of occupation in California was the Rev. William Taylor. In labors he has been more abundant, and as fearless as laborious. His book, as a book of mere incident and adventure, possesses uncommon interest; but as a record of missionary toil and success its interest is immensely increased. The sketches of personal character and death-bed scenes are thrilling." —*Ladies' Repository.*

"The observation and experience recorded abounds with the most pleasing interest, and the scenes are described with much graphic power and felicity."—*Baltimore Sun.*

"This is a graphic description of the labors of a missionary among the most complex, and perhaps most wicked, and at the same time excited and active population in the world. It is a very rich book, and deserves a large sale."—*Zion's Herald.*

"As a religious history, it occupies a new department in Californian literature; and its incidents and triumphant death scenes are of the most interesting character."—*The American Spectator.*

"It is a very entertaining volume, full of adventure, grave and gay, in the streets of a new city, and among a peculiar people."—*New-York Observer.*

"This work is valuable, not merely from its very sincere and sound religious spirit, but from the curious popular traits which it imbodies, and the remarkable insight it affords into the striking and highly attractive peculiarities of the Methodist denomination. We defy any student of human nature, any man gifted with a keen appreciation of remarkable development of character, to read this book without a keen relish. He will find in it many singular developments of the action of religious belief allied to manners, customs, and habits all eminently worthy of study. The straightforward common sense of the author, allied to his faith, has resulted in a shrewd enthusiasm, whose workings are continually manifest, and which enforces our respect for his earnestness and piety, as well as affording rare materials for analysis and reflection. The *naïveté* of the author is often pleasant enough; in some instances we find it truly touching."—*Philadelphia Bulletin.*

"We like the spirit and daring of the author of this book. But few like him live among men. With an undoubted piety, and courage like a lion, he preached Christ at a time, in San Francisco, when Satan reigned about as triumphant as he ever has on any other spot of the cursed earth. The book will be read, and it will do good wherever it is read."—*Buffalo Chr. Advocate.*

"This book is a real contribution to the religious history of that country. For raciness of style it is one of the most readable books that has fallen into our hands."—*Pittsburgh Chr. Adv.*

"The state of society which Mr. Taylor describes is almost anomalous, and his pictures are boldly and clearly drawn"—*New York Evening Post.*

Similar opinions to the foregoing have been given by the Western, Southern, and Richmond Christian Advocates, Christian Advocate and Journal, National Magazine, Methodist Quarterly Review, Harper's Magazine, and many others.

The London Review for April, 1858, devotes nearly four pages to "*Seven Years' Street Preaching in San Francisco,*" from which the following is an extract "The appearance of Mr. Taylor's work on street preaching, at a time when so much attention is turned to this subject, when parochial clergymen, and even bishops, have caught the mantle of Whitefield and the Wesleys, is singularly opportune. And the book itself is so thoroughly good, so deeply interesting, and so replete with wise counsels and examples of what street preaching ought to be, that we cannot but wish for it a wide circulation. The writer tells his story with the simplicity and directness of a child; and the incidents related are of a most unusual and romantic kind. Too much cannot be said in praise of the nervous, plain, vigorous style of the author's preaching. For clearness, directness, and force, the specimens given in this book have never been surpassed."—Pp. 99, 100.

California Life Illustrated.

"Mr. Taylor, as our readers may see by consulting our synopsis of the Quarterlies, is accepted on both sides of the Atlantic, as well as on the shores of the Pacific, as a regular 'pioneer.' The readers of his former work will find the interest aroused by its pages amply sustained in this. Its pictorial illustrations aid in bringing California before us."—*Methodist Quarterly Review.*

"For stirring incidents in missionary life and labors, it is equal to his former work, while a wider field of observation furnishes a still more varied store of useful and curious information in regard to California. It will well repay the reader for the time he may spend on its bright pages. The publishers have done their part well. The book is 12mo., in good style of binding, and printed on fair paper."—*Pittsburgh Advocate.*

"It is a work of more general interest than the author's 'Seven Years' Street Preaching in San Francisco.' It enters more largely into domestic matters, manners, and modes of living. Life in the city, the country, 'the diggings,' mining operations, the success and failures, trials, temptations, and crimes, and all that, fill the book, and attract the reader along its pages with an increasing interest. It is at once instructive and entertaining."—*Richmond Christian Advocate.*

Rev. Dr. Crooks, of New-York, after a careful reading of California Life Illustrated, recorded his judgment as follows: "This is not a volume of mere statistics, but a series of pictures of the many colored life of the Golden State. The author was for seven years engaged as a missionary in San Francisco, and in the discharge of his duties was brought into contact with persons of every class and shade of character. We know of no work which gives so clear an impression of a state of society which is already passing away, but must constitute one of the most remarkable chapters in our nation's history. The narrative is life-like, and incident and sketch follow in such rapid succession, that it is impossible for the reader to feel weary. This book, and the author's '*Young America,*' and '*Seven Years' Street Preaching in San Francisco,*' would make highly entertaining and instructive volumes for Sunday-school libraries. Their graphically described scenes, and fine moral-tone, fit them admirably for the minds of youth."

"Full of interesting and instructive information, abounding in striking incident, this is a book that everybody will be interested in reading. Indeed scarcely anything can be found that will give a more picturesque and striking view of life in California."—*New-York Observer.*

"Mr Taylor has recently published a work entitled *California Life Illustrated,* which is one of the most interesting books we ever read—full of stirring incident. Those who wish to see California life, without the trouble of going thither, can get a better idea, especially of its religious aspects, from this and the former book of Mr. Taylor on the subject, than from any other source conveniently accessible."—*Editor of Christian Advocate and Journal, N. Y.*

"The influx of nations into California, in response to the startling intelligence that its mountains were full of solid gold, opened up a chapter in human history that had never before been witnessed. At first it seemed as if 'the root of all evil,' did indeed shoot into a baneful shade, under which none of the virtues could breathe; but soon Christianity and Gospel missionaries begun to be seen. Among the most active of them was William Taylor, who now, on a return to the Atlantic States, gives to the world a description of what he saw. It is an original, instructive book, full of facts and good food for thought, and as such we heartily commend it."—*Zion's Herald.*

"It is a series of sketches, abounding in interesting and touching incidents of missionary life, dating with the early history of the country, and the great gold excitement of 1849, and up, for several years, illustrating, as with the pencil of a master in his art, the early phases of civil and social life, as they presented themselves, struggling for being and influence amid the conflicting elements of gold mania, fostered by licentiousness and unchecked by the sacred influence of religion, family, and home; containing a striking demonstration of the refining, purifying tendencies of female influence, rendered sanctifying, when pervaded by religion; giving such an insight into the secret workings of the human heart and mind as will be in vain sought for in the books called mental and moral philosophy; withdrawing the vail which ordinarily screens the emotions of the soul, leaving the patient student to look calmly at the very life pulsations of humanity, and grow wise. Statistically the work is of great value to those seeking information concerning the country, with a view to investment or settlement."—*Texas Advocate.*

"The author of this volume is favorably known to many readers by his previous work, in which he relates the experience of seven years' street preaching in San Francisco. He here continues the inartificial but graphic sketches which compose the substance of this volume, and, by his simple narratives, gives a lively illustration of the social condition of California. During his residence in that state he was devoted exclusively to his work as a missionary of the Methodist Church, and, by his fearlessness, zeal, and self-denial, won the confidence of the whole population. He was frequently thrown in contact with gamblers, *chevaliers d' industrie,* and adventurers of every description, but he never shrunk from the administration of faithful rebuke, and in so doing often won the hearts of the most abandoned. His visits to the sick in the hospitals were productive of great good. Unwearied in his exertions, he had succeeded in establishing a system of wholesome religious influences when the great financial crash in San Francisco interrupted his labors, and made it expedient for him to return to this region in order to obtain resources for future action. His book was, accordingly, written in the interests of a good cause, which will commend it to the friends of religious culture in California, while its own intrinsic vivacity and naturalness will well reward the general reader for its perusal."—*Harper's New Monthly Magazine.*

For sale by CARLTON & PORTER, 200 Mulberry-st., N. Y.

www.ingramcontent.com/pod-product-compliance
Lightning Source LLC
LaVergne TN
LVHW020058110826
845151LV00001B/9

* 9 7 8 1 4 2 5 5 4 5 2 9 1 *